AF606868

Guts, Glory, and Gunsmoke

Guts, Glory, and Gunsmoke

Tracing the Lineage of Antique Firearms

Featuring True Accounts from the Civil War, Indian Wars, Winning the West, Building Subways in New York, Revolutionaries in Canada, U.S. Diplomats in the Boxer Rebellion, Rare American Dueling Pistols, & More!

Jerry W. Pitstick

Acclaim Press
MORLEY, MISSOURI

P.O. Box 238
Morley, MO 63767
(573) 472-9800
www.acclaimpress.com

Book & Cover Design: Rodney Atchley

ISBN: 978-1-948901-67-3 / 1-948901-67-6
Library of Congress Control Number: 2020942626

First Printing: 2020
Printed in the United States of America
10 9 8 7 6 5 4 3 2 1

Contents

Acknowledgments

Individuals

"No man is an island," and no one can write a book without help from many others. I am indebted to the following individuals and institutions for their help along the way.

Betty Matthews – Shelby County Historical Society
Bob Boyle – Librarian, Peekskill, New York
Charles Layson – Colt Expert and Dealer
Charles G. Worman – Author and Gun Dealer
Charles Wysong – Descendant of John Wysong
David Allen Comstock – Author
David and Jason Roth – Editor and Publisher, *Blue and Gray Magazine*
David Taylor – Civil War Expert and Dealer
Deborah Magan – Librarian, Shelbyville, Kentucky
Donald Brian – Confederate Gun Expert and Dealer
Donald Rickey Jr. – Museum Curator
Donald White – Historic Photograph Collector
Edward Tyson – Librarian, Nevada County, California
Eloise Hendrickson – Researcher
Justin Runyon – Editor, *Civil War Historian Magazine*
Kandice Liggett Harrison – Editor, *The Gun Report*
Karen Sturm – Museum Curator, Caldwell, Kansas
Kenneth Rymer – Descendant of Charles E. Coffey
Kurt House – Editor, *The Rampant Colt*
M.A. Hoag – Descendant of Colonel John Warner
Mike Speers – Editor, *The Smith & Wesson Journal*
Nancy Rossbacher – Editor, *North South Trader's Civil War*

Norma White – Caldwell, Kansas – Director of the Carnegie Library
Rodney Cook – Researcher
Roy Jinks – Publisher, *The Smith & Wesson Journal*
Roy Marcot – Editor, *The Remington Society of America Journal*
Stuart Mowbray – Editor, *Man At Arms Magazine*
Thomas McNeill Rose III – Descendant of Colonel James McNeill
Vonnie Zullo – Researcher
William R. Scaife – Author

Institutions

Bradley University Library, Peoria, Illinois
Carnegie Library, Caldwell, Kansas
Connecticut Post, Bridgeport, Connecticut
Connecticut Valley Historical Museum, Springfield, Massachusetts
Field Library, Peekskill, New York
Harper's Weekly
Highland Democrat Newspaper, Peekskill, New York
New York Daily Times
Peoria Public Library and City Hall
Searls Historical Society of Nevada County, California
Shelby County, Kentucky Public Library
Springfield Research Service
The New York Times

To my wife, Sharon, for her many years of support and enthusiasm for my gun collecting hobby.

Without her I would not have written this book of short stories about unique people and great pistols.

INTRODUCTION

This book includes twenty exciting, true, and unique stories about the lives and times of the original owners of the fine antique pistols from my personal collection. It is difficult to find an antique pistol whose owner is known, but when you do, there is often an interesting story attached to the pistol and the owner. Once the owner is identified, you must research many sources to unearth the details, which can take several months. I have frequently visited the hometowns of pistol owners to discover local historical societies, libraries, and citizens that often have information on the individual. It usually takes about a year to locate a suitable pistol, perform the research, locate photographs, write the story, and get it published. I wrote my first story in 1998 and my most recent in 2020. This book represents more than twenty years of my writings. I hope you enjoy them all! Please note: many of the photos are very old, so the quality is not as good as we would like but they are the best we have!

Your friend, Jerry

Guts, Glory, and Gunsmoke

Chapter One

Confederate Major Robert McClure, His Colt Pistol, and Sherman's March to the Sea, 1864

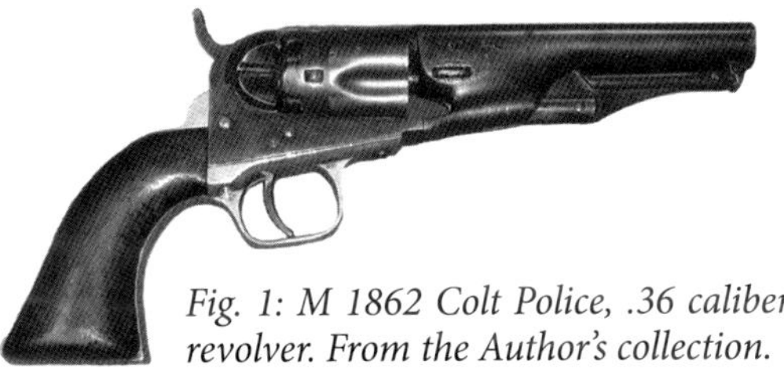

Fig. 1: M 1862 Colt Police, .36 caliber revolver. From the Author's collection.

Major Robert Newton McClure carried this inscribed Colt Model 1862 police revolver (Fig. 1) through a perilous and terrifying journey from the burning of Atlanta in July 1864, General Sherman's epic march to the sea (November-December 1864), and onward to the evacuation of Savannah in late December 1864. This remarkable bisection of the Confederacy by General Sherman was one of the most strategic military maneuvers on American soil, destroying

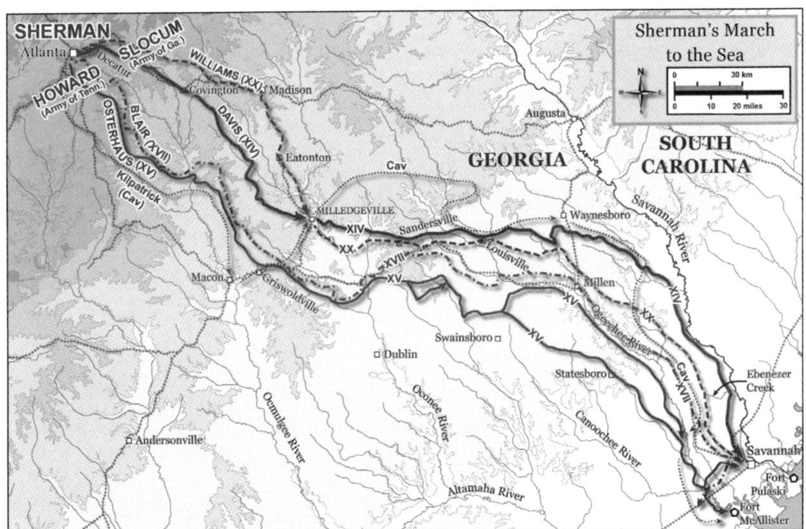

Fig. 2: Map showing General Sherman's March to the Sea, 1864.

about 14,000 square miles of Georgia (Fig. 2 – this is about the size of Lake Erie). Put yourself in Major McClure's boots, as a Confederate soldier and gentleman about to witness in disbelief and horror the destruction of a large part of his home state of Georgia and the complete collapse of the Confederate government, which would soon annihilate the southern way of life as he knew it.

He was an adventurous, young, handsome, and rich southern gentleman and plantation owner eager for the glory of war and achievement of military rank. These feelings were shared by most young, southern aristocrats, who didn't necessarily consider the potential consequences of war, i.e. falling ill, losing the war, becoming a prisoner, or being wounded in battle, with permanent disability or even death! Perhaps this was a high price to pay for ambitious yearnings and upward mobility!

But, I am getting ahead of myself and must first explain my good fortune in finding this revolver and having the luck (or wisdom) to purchase it. For many years I have specialized in collecting inscribed Civil War pistols and revolvers, and inscribed "Win The West" type revolvers. I have never been a dealer, but rather a collector who enjoys displaying and researching the often very interesting life stories attached to the owners of inscribed or identified guns. These relics represent important pieces of American history that have survived over 150 years.

For the first time in May of 2003, my wife, Sharon, and I went to the Nashville Civil War Show (Tennessee) held at the Nashville Fairgrounds. This show far exceeded our expectations, but in talking to various dealers I learned that this was probably the number one Civil War show in the country, especially in attracting the greatest quantity and quality of Confederate pieces.

Having lived in the North my entire life, I had relatively little exposure to Confederate revolvers of any kind, especially those revolvers actually manufactured in the South, i.e., Spiller & Burr, Leech & Rigdon, J.H. Dance & Brothers, Griswold & Gunnison, Schneider & Glassick, and Tucker & Sherrard, etc. Of course, I had always wanted to include one of these in my collection, but ran into two major obstacles: 1) Lack of specimens to examine and, 2) Some prices that were beyond my means and rising fast.

Over a two-day period, I looked at every Confederate revolver at the show including not only southern manufactured guns, but also northern guns with Confederate provenance. The number of such guns at the show was approximately thirty. The price range of the southern manufactured guns ran from $11,000 for a heavily used Spiller & Burr with some replaced parts, to $100,000 for an inscribed J.H. Dance & Brothers revolver in good condition and with credible provenance. The latter was the only inscribed southern manufactured revolver at the show, and was far beyond my budget. So, I concentrated on northern manufactured guns with a guarantee of authenticity and proof of ownership by a Confederate soldier, preferably an officer.

Mr. Donald Bryan of Roanoke, Virginia, an outstanding dealer with a sterling reputation, was there and it so happened we had conducted business ten years earlier at the Richmond Civil War Show (Virginia). Don had an 1862 Colt police with a 4-1/2" barrel manufactured in 1863, Serial #16464, inscribed "Major R.N. McClure, 11th Regt. Ga. Mil. Cav. 1864" on the brass backstrap (Fig. 2A). Don's letter to me states this gun and its inscription were unconditionally guaranteed to be authentic. In addition, he had portions of McClure's military records and a muster roll of Company I, 52nd Regiment, Georgia Volunteer Infantry, Army of Tennessee, C.S.A., Dawson County, Georgia.

These records indicated that Robert Newton McClure had the following condensed military career: 1st Lieutenant, March

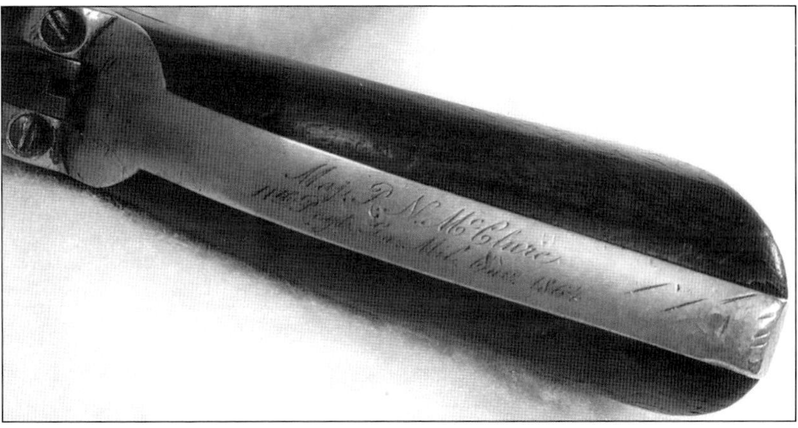

Fig. 2A: 1862 Colt Police inscribed "Major R.N. McClure 11th Regt. GA. Mil. Cav. 1864." Photography by Scott Guyton.

4, 1862; contracted typhoid fever at Big Creek Gap, Tennessee, May 1862. Typhoid settled in both legs and erysipelas in his head caused loss of hearing in his left ear. He resigned February 2, 1863, re-enlisted July 7, 1863 and was elevated to Captain of Co. D., 4th Battalion, Georgia State Guards Cavalry; Major of 11th Regiment Company S, Georgia Militia Cavalry in May 1863; surrendered at Kingston, Georgia, May 12, 1865.[1]

Fig. 3: Major Robert N. McClure, 11th Regiment Georgia Militia Cavalry, 1864. Illustration by Artist, Sharon Pitstick.

In 1860, Robert Newton McClure had a sizeable net worth of $4,900 - $10,500[2] inherited from his father, Robert B. McClure, an affluent planter worth over $24,000. This made twenty-five year-old Robert N. McClure a rather wealthy and influential young gentleman. In 1860, land was available for about one to three dollars per acre in northwest Georgia. This means that he was capable of purchasing a plantation in excess of 2,000 acres, construct a fine home on the property, and still have sufficient funds left safely in the bank.

Robert had married the very desirable and socially prominent Emma Hockenhull in 1855 when he was twenty years old and Emma was eighteen years of age. In 1862, with three wonderful children and being only twenty-seven years old, Robert N. McClure was seeking fame, glory, and military rank (Fig. 3). This drive led Robert to enlist. Unfortunately for Lieutenant McClure, on May 15, 1862 at Big Creek Gap, Tennessee, he fell ill with typhoid fever and was captured by the Union forces. He had no medical treatment for three months. The typhoid settled in both legs, which swelled up, producing permanent varicose veins in addition to his loss of hearing. He was forced to resign in February 1863, because he was no longer able to perform the duties of a soldier.[3] He was paid $450 for his time and efforts in the Confederate Army.[4]

Upset and sick that he was missing this glorious war, he healed enough in six months to ride again, but the Confederate Army would not allow a man so injured to re-enlist. Lieutenant R.N. McClure was well connected and socially popular in Georgia, as his ancestors comprised a prominent family line, one of who served in the Revolutionary War (John McClure Jr., born April 15, 1760).[6]

Lieutenant McClure knew Georgia Governor Joe Brown socially, and persuaded him to commission him Captain of Co. D., 4th Battalion Georgia State Guards, Cavalry, in July 1863. This was Robert McClure's opportunity to get back into the fray and perhaps win fame and glory.

At the beginning of the Civil War, Georgia ranked third among the Confederate States in manpower resources, with an arms-bearing population exceeding 120,000 white males, aged sixteen to sixty. After many of these men volunteered for Confederate service, those who remained behind became the prize in a fierce struggle between Georgia Governor Joseph Brown (Fig.4) and Confederate President Jefferson Davis. This conflict premised on both states' rights ideology and power politics, which raged throughout the war. Consequently, the state's militia underwent frequent transformation as the Confederate central government increasingly found ways to pull Georgia's available men into the national army. Nonetheless, as late as 1864 Governor Brown fielded a sizable militia division – about 3,000 men, principally composed of boys, old men and men already discharged from the Confederate Army. Its men participated in both the Atlanta and Savan-

Fig. 4: Georgia Governor Joe Brown. Georgia Archives.

nah campaigns. Known by this time as "Joe Brown's Pets," the Georgia Militia remained under the Governor's control until he surrendered them to invading union forces in May 1865.[5]

Fig. 4A: Major General Gustavus W. Smith. Library of Congress.

This new Georgia force of 1864 was designated the First Division, Georgia Militia, organized into four brigades of three regiments each, plus cavalry and artillery units, and would be commanded after June 1, 1864 by Major General Gustavus W. Smith (Fig. 4A). Under Smith, the militia would have its most significant impact on history, experiencing combat for the first time and providing valuable service during the final phase of the Atlanta campaign.[5] On July 24, 1864, during the siege of Atlanta at Fort Hood, one of the Confederate cannons was fired directly over Major McClure's head. The concussion was so severe that his hearing was again injured. His hearing would continue to worsen until by July 1902, at sixty-seven years old, he would be totally deaf.[7] The militia would constitute the major Confederate force at the battles of Griswoldville and Honey Hill during General William Tecumseh Sherman's March to the Sea, and would play a prominent role in the siege and evacuation of Savannah.

In researching the Georgia Militia, the most voluminous single source is a collection from the Adjutant General's office of the original rosters of the militia taken at its reorganization as the 1st Division, Georgia Militia. These are found at the Georgia Department of Archives and History in Atlanta, where access to them is limited because of their delicate state of preservation. Our subject soldier, Robert McClure, is listed on page 273 of the book, *Joe Brown's Pets—The Georgia Militia 1861-1865*, by historians William R. Scaife and William Harris Bragg. This book was published in 2004 by Mercer University Press, 1400 Coleman Avenue,

Macon, Georgia, and is must reading for descendants of Georgia Militia soldiers, because their history was not compiled previously. Captain Robert N. McClure served Governor Joe Brown faithfully and well, though in routine assignments, for the ten months from July 1863 to May 1864. He was rewarded accordingly by his promotion to Major 11th Regiment, Co. S Georgia Militia Cavalry, where he would find himself involved in some of the most glorious, though unfortunate, war assignments in Georgia history.

Twenty-nine-year-old Major McClure apparently celebrated his promotion by purchasing an expensive and hard to get memento — a Model 1862 Colt Police, 6-Shot Revolver with personal engraving on the back strap. Perhaps it was not so much a memento as it was his recognition that he would play a part in the hellish assignment of protecting Atlanta from General William Sherman, who was pressing southward from Chattanooga with the intention of taking Atlanta. In other words, Major McClure was going to need maximum firepower, which was becoming scarce in the south.

Cavalry General Joseph Wheeler (Fig. 5) of the Regular Confederate Army was to continue his role of reconnaissance and communication between Major General John Bell Hood, commander of the Confederate forces during the siege of Atlanta, and Major General Gustavus Smith, a former Confederate General and now Commander of the 1st Division, Georgia Militia.[5] Most cavalry units of the Georgia Militia were placed under the command of General Wheeler from the Atlanta campaign to the evacuation of Savannah, including General Sherman's march to the Atlantic Ocean.

Fig. 5: Confederate Cavalry General Joe Wheeler. Hargett Rare Book and Manuscript Library, University of Georgia.

General Smith named his

staff officers on June 11, 1864, among them General Robert Toombs, former member of President Davis' cabinet, Colonel Joseph Claghorn (former Commander of Savannah's Artillery), and aide-de-camp Linton Stephens, half brother of confederate Vice-President Alexander H. Stephens. Also, in June 1864 two militia brigadiers were selected – Brigadier General R.W. Carswell for the 1st Brigade, Brigadier General P.J. Philips for the 2nd Brigade. Also organized was an independent Artillery Battalion under Colonel Carey W. Styles.

On June 24, General Smith was ordered to lead his militia troops in support of General William H. "Red" Jackson's Cavalry Division, posted on the extreme left of the Confederate Army in defense of Atlanta against Sherman's ever-advancing forces. At this time our Major McClure and his Militia Cavalry were under the temporary command of General "Red" Jackson's Cavalry Division. Desperate fighting was everywhere because of Atlanta's importance. On July 5th, Smith was ordered to withdraw and fall back into the Chattahoochee River line because Sherman could not be stopped.

General Smith asserted that the militiamen had executed the support and delaying assignment thoroughly in their first major combat role. The Militia's commendable performance prompted Confederate General Joseph E. Johnston to write a letter of commendation to "His Excellency Joseph E. Brown, Governor." "According to all accounts the Militia's conduct in the presence of the enemy was firm and credible."

After a July 22nd engagement of the Militia, it was reported by Confederate General Hardee (Fig. 6) that, "…within ten minutes the effective fire power of the enemy was silenced in our front, and after this the Federals only occasionally would stick their heads above the parapet. … The Officers and Militiamen behaved admirably, every movement was promptly and accurately made."[5]

A few days after the July 22nd Battle of Atlanta was lost by Confederate General John Bell Hood (Fig. 7), along with 40,000 Confederate troops to Union General William T. Sherman (who lost 30,000 troops), General Smith described his militia troops in reporting to General Hood on their part in the Atlanta campaign: "The Militia, although but poorly armed and trained — all

Fig. 6: Confederate General William Hardee. U. S. Military History Institute.

Fig. 7: Confederate Major General John Bell Hood. Hargett Rare Book and Manuscript Library, University of Georgia.

performed well every service required during a dangerous campaign. They have only been in service about one hundred days during at least fifty of which they have been under close fire of the enemy mostly night and day. They have always shown a willing spirit, whether in camp, on the march, working on fortifications, guarding trenches, or upon the open battlefield. They have done good and substantial service in the cause of their country and the State of Georgia."

On September 1, 1864, fallen Atlanta was evacuated of citizens per General Sherman's demands so that Sherman's men could torch the city's commercial structures and buildings that could serve the Confederacy. Atlanta was known as the "Workshop of the South." Sherman (Fig. 8) did not directly intend the torching of residences, but in the conflagration, the fire got out of control and burned down about 4,500 homes in the city.[9]

October 3rd, General G.W. Smith was ordered to reassemble the militia and his cavalry, including Major Robert McClure, in the vicinity of Macon — as soon as transportation could be arranged — to make a demonstration against Sherman's forces in Atlanta to discourage Sherman from taking Macon, which now was the safest place for the Georgia State Government.

Fig. 8: Union General William T. Sherman. Hargett Rare Book and Manuscript Library, University of Georgia.

Of course, the Confederate High Command had no idea what Sherman's next move might be. Would Sherman go north to pursue General Hood's army, which may retake Nashville and threaten Kentucky and even Ohio? Or would Sherman move south to take Macon, or east to take Augusta? History now tells us, Sherman would instead do the unthinkable — march southeast in between fortified Macon and heavily guarded Augusta, right through the fertile "Garden of Georgia" burning most structures, impounding all livestock and food in the direction of Savannah and the sea. Sherman's men were required to forage 60,000 lbs. per day or risk starvation!

The great march to the sea began with a spectacular and memorable scene on November 16, 1864. That day Sherman sat astride his horse, and with stern manner and exultant heart, watched his army, some 60,000 battle hardened veterans and 5,500 cavalry, stripped down for action, swing past shouting and singing, "Glory, Glory, Hallelujah!" They were on the way to the sea — perhaps even to achieve a junction with Grant and his army! Forty thousand of Sherman's men were to remain behind to hold Atlanta and go north toward Chattanooga to protect Sherman's supply line. A southern gentleman said, "Hell has laid an egg and it hatched in Atlanta."

Sherman's intrepid decision to transfer his Army to Savannah and the seacoast made many northerners nervous. General Thomas, commander of one wing of Sherman's army, advised against it. Lincoln was apprehensive. General Grant thought the safest course would be to destroy Hood's army first. Sherman was determined that he would not let himself be maneuvered out of Georgia, and finally through perseverance received Grant's and Lincoln's approval.[9]

On November 22, Sherman's Cavalry, under General Slocum, occupied Milledgeville, the capital of Georgia, which is about one-third of the way from Atlanta to Savannah. Ironically, General Sherman occupied the executive mansion of Governor Joe Brown to protect its contents. Milledgeville lies about forty-five miles northeast of Macon. The only organized opposition at Milledgeville had come from Confederate cavalry under General Joe Wheeler and Major Robert McClure and his militia cavalry. This threat to Macon caused the Confederate forces under Lt. General William Hardee and the militia under General Smith to head northeast from Macon about ten miles to the vicinity and east of Griswoldville, where they dug in trenches and fortified the line to protect Macon and harass Sherman. As Sherman proceeded southeastward, all railroad tracks were destroyed by bending the rails around trees; these sights became known as 'Sherman's Neckties.'

Griswoldville was a small industrial village that had a railroad depot, cotton gin, furniture factory, brick plant, and pistol works that manufactured Griswold and Gunnison Revolvers. On November 24, 1864, Union forces struck the village destroying everything including the church and a locomotive with thirteen rail cars. Every structure was destroyed except for Sam Griswold's house and one other residence. General Sherman was acclaimed as the "Father of Total War."

Then a large Federal force under Colonel Robert Catterson struck the main body of General Smith's militia and their dug-in fortifications. The overwhelming Federal might crushed the militia, who fought valiantly but to no avail. After dark, Militia General Philips withdrew his battered forces west of Griswoldville, where trains carried his men to the safety of the fortification surrounding Macon. The Confederates lost about 522 men, or 22% of approximately 2,300 participating. The Federals reported 94 lost, or about 3% of the approximately 3,000 present. Whatever the reason, the militiamen had no shame regarding their loss, since their bravery was now unquestioned.[5]

By December 3, 1864, Sherman's forces freed the Union prisoners being held in Camp Lawton, the Confederate prison at Millen, Georgia, just seventy miles south of Augusta. Northern newspapers soon carried ghastly pictures of emaciated prisoners, just released, and of the clay mounds in the prison pen under which they had crawled to find refuge from burning sun and beating rains. Some unburied corpses lay near the graves of 700 dead.

After Millen, Sherman gave orders to march on toward Savannah. As they approached the sea, the country became lower and more marshy; marching was difficult and there was very little food to forage. Additional obstructions were presented by trees chopped down across the roads, particularly near creek, swamp and causeway crossings, but Sherman's workers under West Point Engineer, Col. Orlando Poe, rapidly removed them. Sherman also encountered the first use of land mines, called torpedoes, which southern soldiers had buried in and around the crude roadways.

By the time the Army was within fifteen miles of Savannah, opposition thickened and so did the barricades, while supplies became dangerously shorter.[9] Starvation became a real possibility if Sherman was stopped for long. Sherman did not know the strength of southern forces at Fort McAllister, just south of Savannah, but he knew he must take the fort quickly to avoid starvation and possible defeat. Fortunately for Sherman, troop strength was weak at Fort McAllister, and his battle-hardened men took control of the fort and captured the rebel soldiers in less than an hour of battle.

Meanwhile, General Smith and his Georgia Militia troops arrived in Savannah from Macon late on the evening of November 29th, exhausted from their several days of circuitous constant travel and hoping to get some much needed rest. Instead of rest, General Smith found orders awaiting him to take his militiamen north to the vicinity of Grahamville Railroad Station in South Carolina. His mission was to protect the Charleston and Savannah Railroad from destruction by the Union forces at Fort Foster on the South Carolina seacoast. This railroad line was the only means of escape for General Hardee's 10,000-man army and citizens, should Savannah fall to Sherman.

The South Carolina Cavalry under Colonel Charles J. Colcock, including Major McClure's Georgia Militia Cavalry, and General Smith chose a slightly elevated place known as Honey Hill as a suitable defensive line to protect the railroad. In addition, Colonel

Colcock devised a delaying plan for the advancing Federal troops. He ordered Major McClure to take his Cavalry and set ablaze the fields across which the Federals were marching. The dry broom grass, fanned by a brisk breeze, flamed up and sped toward the Union troops at head height, temporarily halting their progress.

During Colcock's three-hour delaying action, the line at Honey Hill was fully prepared; to the left were the First Brigade, Georgia Militia commanded by Colonel James Willis; the Georgia State line under Lieutenant Colonel James Wilson; and the Augusta Battalion, Major George T. Jackson commanding. Beyond them to the left, some dismounted cavalry and one cannon guarded the flank. In the center of the works, Major F.W.C. Cook's Athens, Georgia Battalion was posted, supported by seven artillery pieces. The confederate line extended thinly beyond the fortifications over a mile to the right, held by a few men and one artillery piece; the ground to their front being practically impassable. To the rear, the 47th Georgia was held in reserve at the direction of General Smith.

Colonel Colcock wrote describing the battle, "At half past ten o'clock, the head of the Federal column appeared around a bend in the road about 120 yards from our breastworks apparently unconscious of its existence and doubtless anticipating a pleasant morning march to the railroad. They little knew what was waiting for them. A wild Confederate cheer — a volley from a thousand unerring rifles and six Napoleon guns and the whole of their column went down to their long sleep of death. Not a human soul was left but one mangled mass of dead and dying crimsoned with their life's blood."[9]

For their part, the Georgia Militiamen held firm, as the Union advance emerged from the cover of the swamp and approached the Confederate works. Behind "Joe Brown's Pets" one of the officers exhorted them, "Keep cool, boys, and when they come out of the swamp, shoot low." The end result, the Confederate line was never reached or penetrated at any point.

By dusk, Federal ambulance wagons had arrived; the Federal force began falling back to the Bolen Church. There the pews were removed and converted into a hospital to accommodate the large number of wounded. In the church yard "trenches yawned to receive amputated limbs." The Federals reported losses of eighty-nine killed, 629 wounded, and twenty-eight missing for a total of 746 men out of a force of 5,500. The Confederates reported losses of eight killed and forty-two wounded out of a force of 1,400. Predictably, General Smith recounted the events at Honey Hill with considerably more enthusiasm than he had described the Griswoldville battle the previous week.

By December 10th, the 10,000 Confederates under General Hardee, including the Georgia militiamen under General Smith and Major Robert McClure's cavalry, were forced back into the fortifications of Savannah because of Sherman's relentless advance. Here, General William Hardee was protected by swamps, creeks, and flooded rice fields. Union Cavalry General Slocum's columns were supported by Davis' corps on the left on the Savannah River, while other troops on the right extended the siege line toward the sea, where Union Admiral Dahlgren's fleet was awaiting its opportunity to support Sherman in the siege of Savannah.

On December 17th, Sherman demanded Hardee's surrender with his 10,000 men, but Hardee had decided against surrender and was planning an escape by way of the one causeway still open across the Savannah River. Realizing that he could no longer delay evacuation, he completed the movement on the 20th. At daybreak next day, Sherman's 20th Corps entered Savannah. Sherman himself followed on the 22nd, and he wrote, "I was very much disappointed that Hardee escaped with his garrison and had to content myself with the material fruits of victory without the cost of life which would have attended a general assault."[9] During Sherman's march about four billion dollars worth of slaves were liberated.

General Sherman, now forty-four years old, comprehended the full import of this dramatic triumph and electrified the nation with his famous telegram from Savannah, December 22nd, "His Excellency, President Lincoln, I beg to present you, as a Christmas gift, the City of Savannah, with 150 heavy guns and plenty of ammunition, and also about 25,000 bales of cotton." Military experts think it remarkable that Sherman and his men took only thirty-seven (37) days to reach the sea under difficult and frightening conditions. The whole army averaged about seven miles a day and occasionally advanced as much as fifteen miles in a day – a genuine feat.

Obviously, "Joe Brown's Pets," the Militiamen (Fig. 9), including our Major Robert Newton McClure, played a significant part in these last months of the Civil War. Major McClure did achieve a certain amount of fame both locally and throughout the state of Georgia, and a bit of glory, eventually reaching the rank of Major and cited for distinguished service by the Confederate Secretary of War. Major McClure received a $50 per year pension under Claim #985360 for the injuries he sustained in the line of duty, but all participants now understood that, "War Is Hell."

Fig. 9: Georgia Militiaman before and after the war. Charles Henry Smith, "Bill Arps Peace Papers," (1873).

Major McClure surrendered at Kingston, Georgia, on May 12, 1865. In later years he was usually known as "Major" or Major Newton. He and his wife, Emma (Fig. 10), had nine children — five girls and four boys, plus they adopted their orphaned neph-

Fig. 10: Robert and Emma McClure with daughter in front of their home, ca. 1888. Courtesy Georgia Archives.

Fig. 11: Robert McClure's General Store in Dawson County, Georgia, ca. 1890. Courtesy Georgia Archives.

ew, Robert Newton Cantrell. Robert McClure was a planter and general store owner in Dawsonville (Fig 11), Lumpkin County, until 1904 when he retired to the adjacent City of Gainesville, Hall County, Georgia. Major McClure died November 19, 1915 (Fig. 12) at the age of eighty, and his wife, Emma, died March 25, 1918 at eighty-one years of age.[10] Robert and Emma were happily married for sixty years, which was extremely rare in the eighteen hundreds and early nineteen hundreds. They are buried side by side in Woodlawn Cemetery in Gainesville, Georgia.

After Sherman's great accomplishments, Congress wanted to make General Sherman equal to General Grant, but Sherman declined the offer, preferring Grant's trust and friendship over his own elevation. Sherman also declined opportunities to run for President of the United States. After the war, General Sherman made his living on

Fig. 12: Robert N. McClure's grave stone at Woodlawn Cemetery, Gainesville, Georgia. Courtesy Woodlawn Cemetery, Gainesville, Georgia.

lecture tours and personal appearances at major events. He gave much of his money to old soldiers in need. All of these facts serve to demonstrate Sherman's strength and character.

References

1. Muster Role of Co. I, 52nd Regiment, Georgia Volunteer Infantry, Army of Tennessee, C.S.A., Dawson County, Georgia.
2. Source – 1860 U.S. Census.
3. Letter of Resignation from Lt. Robert N. McClure to Secretary of War, James A. Seddon, Richmond, Virginia.
4. Source – Confederate States of America, Commissioner of Pensions, State of Georgia, City of Atlanta.
5. Scaife, William R. and Bragg, William Harris, "Joe Brown's Pets" The Georgia Militia 1861-1865, Mercer University Press, Macon, Georgia, 2004.
6. McClure, Donald A., A Living Descendent and Author of, *The North Georgia Pioneer Families of Robert B. McClure / Mary Keith and John Cantrell / Martha Porter*, 1995, revised 1997.
7. Zullo, Vonnie S. The Horse Soldier Research Service, Gettysburg, Pennsylvania.
8. DeLoach, Gail, Georgia Division of Archives and History, Visual Archivist.
9. Hattaway, Herman, *Shades of Blue and Gray, The War For The Union* Sherman's March To The Sea.
10. University Microfilms, Inc., Ann Arbor, Michigan for Book, *The McClure Family,* published 1905, by The Reverend James Alexander McClure.

Chapter Two

Captain James Ruby Brawling Indian Fighter & Civil War Hero

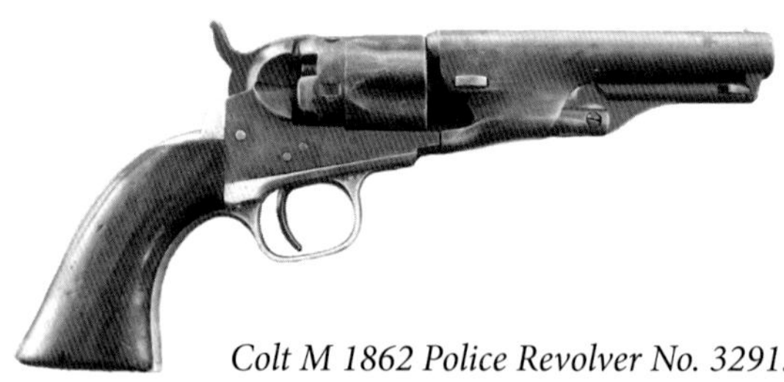

Colt M 1862 Police Revolver No. 3291.

On May 29, 2009, I was installing my gun display at the Annual Ohio Gun Collectors Show in Wilmington, Ohio. Chuck Worman, one of my favorite gun dealers, stopped by and, in the course of our conversation, I told him I was still interested in acquiring Civil War inscribed pistols. Chuck reached into his satchel and pulled out a Colt M-1862 Police Revolver #3291, manufactured in 1861, with the following inscription on the backstrap: "Pres'd To J. N. Ruby by his Co. E 2nd Wis. Reg't Sept. 21, 1861."

This type of inscribed gun is "right down my alley," so we negotiated a price and the Colt was mine along with the basic military service paperwork that reads:

Inscription on the backstrap of Captain James Ruby's Colt Police. "Pres'd to J.N. Ruby by his Co. E 2nd Wis. Reg't Sept. 21, 1861."

> "August 1855 to 1860 Private Troop E, 1st U.S. Cavalry — Fought Indian Wars against Apache, Navajo, Yakima, Cayuse Takelma, Shasta, Nez Perce, Chetco, Tututni, Creeks and other tribes. He was in dozens of skirmishes and two major battles — The Rogue River War and The Battle of Hungry Hill in Oregon Territory. In the latter, troops had to retire from battle after fighting for two days, with twenty-six soldiers killed or wounded. James Nelson Ruby was discharged August 21, 1860, having completed his five years of service.
>
> "Eight months later, he enlisted in the Union Army; April 20, 1861 to November 27, 1861 – Co. E 2nd Regiment Wisconsin Volunteer Infantry, commenced as a Private, then Sergeant Major on July 16, 1861, then promoted to 2nd Lieutenant on August 8, 1861. To honor 2nd Lieutenant James N. Ruby, his company gave him an inscribed Colt. Lieutenant Ruby must have been an esteemed officer to have been presented with such an expensive and cherished gift from the troops. June 28, 1862 to December 1862 – Co. K, 20th Regiment Wisconsin Volunteer Infantry. December 1862 to September 8, 1863 – Co. B 34th Regiment Wisconsin Cavalry. Ruby was promoted to Captain upon joining this unit, having had extensive experience in the 1st U. S. Cavalry."

In total, Ruby served honorably and heroically for twenty-one months in the Civil War and five years in Indian wars.

In 1855, Ruby was handsome, single, twenty-two years old, 6' tall, 140 pounds with gray eyes, dark brown hair, and a dark complexion, and he was ready for achievement, adventure, and action.

Upon his arrival in Oregon Territory, he found all hell breaking loose! The discovery of gold in southwestern Oregon led to wars between whites and Native Americans. THE GOLD RUSH WAS ON! In spite of the promises of two superintendents of Indian Affairs, Mr. Dart and Mr. Palmer, white people were flooding into the area. Dispossession ruled and new and fatal diseases attacked the Indians. The miners drove numerous Indian tribes from their villages, many of which were located on old stream terraces that were prime locations for gold placer deposits.

The hungry newcomers hunted the game, decimating deer and elk populations. In 1854, the territorial legislature prohibited the sale of ammunition or guns to Indians, deepening their disadvantage and anger.

Some vigilante-type whites banded together in the mid-1850s as "Exterminators" to kill Native Americans in Oregon Territory, including those camped near Table Rock. Mining debris poured down the rivers, salmon runs diminished, and the eels died. Crayfish, mussels, and trout choked on the flood of mud, and starvation threatened the native population.

The mining districts in the Rogue River country caused major ecological disruption, unraveling nature's ways and long-established human subsistence activities. The "Exterminators" perpetrated massacres, killing about one hundred Indians without mercy, including women and children. After these outrageous abuses, it comes as an ironic surprise that the *New York Daily Times* of October 21, 1855 broke the following story:

> ***"Indian Outbreak In Rogue River Valley, Twenty-Two White Families Murdered."***
>
> "Governor Curry, this morning, called for help from the U.S. Army, 1st U.S. Cavalry and as many volunteers as possible. Of course, James Ruby was deeply involved in this activity as he was appointed to aid in volunteer recruitment. One battalion is to be raised in the north portion of the Oregon Territory and the other in the south, both to act in concert in avenging and exterminating the Indian vermin from the mountains and hills that surround our valleys. Oregon seeks to have 2,000 troops in the field contending against savages who are acting together as a grand war party of extermination against the whites."

The Rogue River Indian War re-ignited from its first rumblings of 1853-54. By late 1855, the homes of settlers and miners homes were now fortified; all but essential business was suspended and every available man started out to hunt down and murder Indians.

Finally, the U.S. Army developed sufficient forces to mount a campaign in 1855 and 1856 to destroy the Indians' ability to resist. This campaign was led by the 1st U.S. Cavalry, headquartered at Fort Tejon, California, commanded by the following officers: Colonel Fauntleroy, Major Chandler, Captain Wright, and 1st Lieutenant Steptoe.

Lady Oscharwausha, the last of the Rogue River Indians in the valley.

It was believed that white women were prisoners among the Indians near Table Rock, a rumor that began from vague reports of the capture of two white girls near Klamath Lake. The U.S. Army and 1st U.S. Cavalry attacked the Indian stronghold along with several hundred volunteer soldiers, with the result of both freeing the captive women and destroying hundreds of Native Americans — much of the fighting being hand-to-hand brawling in close quarters! Upon going into the Indian camp, James Ruby reported many wounded, and the Indians were burning their dead as if fearful the bodies would fall into the hands of the enemy.

The famous Table Rock in the Oregon Indian Territory.

Colonel Fauntleroy was met by Chief Jo, who told him their hearts were sick of war and that they would meet him seven days thereafter at Table Rock when they would give up their arms, make a treaty of peace, and place themselves under the protection of the Indian superintendent, Joel Palmer.

Indian Superintendent Joel Palmer.

Following the Rogue River battles and the Battle of Hungry Hill, Mr. Palmer negotiated treaties with all the Native American Tribes living in the troubled areas of the Oregon Territory. Vanquished by the combined operations of the Oregon volunteers and Army regulars, the Indians of the Rogue and Umpqua Valleys and the southwestern Oregon coast were then removed to the Siletz and Grand Ronde Reservations.

Forced marches through winter snows or over rocky headlands and through the sand dunes of Coastal Oregon became "trails of tears" for hundreds of Indians driven to the distant reservations. Other survivors were herded aboard the ship, *Columbia*, a side-wheel steamer, which removed them from Port Orford to the Columbia and lower Willamette River areas. Then they had to walk the muddy trail to the reservations.

The river at the bottom of this gorge provided rich fishing for the Indian population.

Few Indians wanted to engage in the backbreaking agrarian lifestyle or give up their fishing, hunting, and gathering way of life, but it was forced upon them. More immigrants arrived every fall, settlements spread, pioneer cabins lined the shores of the rich fishing gorge, and worse, more gold was discovered.

The late 1850s was a wrenching time of transition. Steadily the Indian numbers diminished, their food sources destroyed, and their lands appropriated. These were devastating times, indeed, for the region's native peoples.

In less than nine months after James Ruby's discharge from the 1st U.S. Cavalry on August 20, 1860, he gradually developed two opinions:

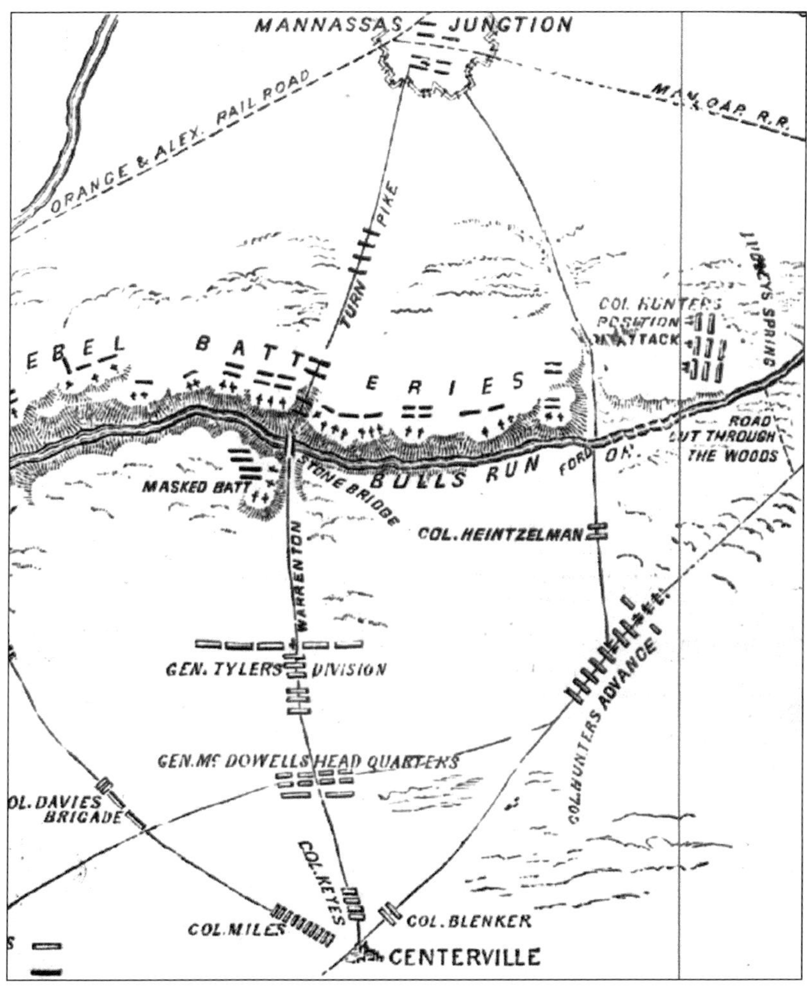

Map of Centerville and Manassas Junction.

1. He was weary of paint contracting and a relatively dull civilian life.
2. He saw his five years of regular U.S. Army service training as an opportunity to advance quickly up the ranks of the newly formed Union Army, after Lincoln's call for 75,000 troops to put down the "rebellion."

Consequently, on April 20, 1861 at Oshkosh, Wisconsin, Ruby enrolled in Captain Bouck's Co. E, 2nd Regiment, Wisconsin Infantry. He was mustered in on June 11, 1861, at Madison, Wisconsin. Little did Ruby comprehend that on this date he was less than six weeks away from murderous fighting and a disastrous Union loss at the Battle of Bull Run, also known as First Manassas. The battle took place on July 21, 1861. Co. E was attached to and part of Colonel William T. Sherman's "Iron Brigade," General Daniel Tyler's Division and Commanding General Irwin McDowell's Army of Northeast Virginia. Sherman and Tyler enjoyed excellent reputations and experience in the *old* U.S. Army, as did James N. Ruby, but McDowell was inexperienced, and it would soon show.

The "Iron Brigade" consisted of the 2nd, 6th, and 7th Wisconsin, 19th Indiana, and 24th Michigan Regiments, which amounted to about 4,000 to 5,000 of the 20,000 Union soldiers engaged in the first battle of Bull Run. The Confederates strength was near 17,000 soldiers. At Bull Run the 2nd Regiment, including our James Ruby, bore the brunt of a determined attack by Confederate General "Stonewall" Jackson's entire division on the Warrenton Pike until the Union brigade could be moved into position and the enemy repulsed. It was here that Ruby's staunch determination earned him a hero's status, the admiration of his men, and the soon-to-come promotion to 2nd Lieutenant. The brigade held the line of battle until the Union Army had passed to safety on the road to Centreville. Later in the day there was a vicious engagement on the Warrenton and Sudley Roads.

Only experienced military men expected the lovely countryside of northern Virginia to host this first major battle of the Civil War on July 21, 1861. The reason for this is that Manassas Junction, located in between Washington, D.C., and Richmond, Virginia, controlled the crucial East-West Manassas Gap Railroad and the North-South Orange and Alexandria Railroad. Marching southward from Centreville, Commanding Union General Irwin McDowell's leading column, under General David Hunter, moved along the Warrenton Turnpike toward the deep strong current of the Bull Run River. Then Hunter and General Heintzelman turned off the pike in a wide sweep upstream to cross Bull Run at Sudley Ford northwest of the infamous stone bridge. Meanwhile, General Dan Tyler's Division—including James Ruby—was to move straight along the turnpike to the stone bridge, the only bridge in the area across Bull Run. All other possible troop crossings would have to take place at various shallow fords.

Nathan G. Evans (Massachusetts Commandery Military Order of the Loyal Legion and US Army Military History Institute).

Ironically, General Beauregard and General McDowell had mirror image battle strategies in that both generals intended to go around the right flank of the opposition, creating laughably a do-si-do square dance effect.

General P.G.T. Beauregard, who was chiefly responsible for the Confederate positions, was taken by surprise, having expected an attack on his right. He had put the bulk of his forces in place to cover the lower, more southeasterly fords. His left flank at the northwesterly stone bridge had nothing more than the half brigade commanded by Colonel Nathan G. Evans to cover it. Luckily, for the Confederates, Nathan was a classic southern brawler, outnumbered ten to one and without additional support, save his 400 men, his instinct was to attack rather than cower.

McDowell's plan worked wonderfully at first. Tyler did his job well and Hunter with Heintzelman, had crossed Bull Run to the northwest at Sudley Ford without opposition. When the Federal masses swept south from the ford, Evans was hard pressed to hold them at Matthews Hill west of the stone bridge, but hold them he did! Meanwhile, the Rebel high command reacted quickly and sent reinforcements to their endangered left flank, with Rebel brigades under General Barnard Bee. Then Colonel Francis Bartow arrived, allowing the Confederates to deliver a staggering counterattack. However, Yankee pressure forced them south from Matthews Hill toward Henry Hill. Now General Thomas J. Jackson arrived to become a "Stone Wall" and even more Rebel reinforcements stabilized the line though charges and countercharges continued for over two hours.

As new fresh forces arrived on both sides, the lines grew longer, extending to the southwest as more soldiers reached the scene. Yankee General Tyler, commander of our man, James Ruby, finally started to cross Bull Run at the bridge while the fight to control nearby Henry Hill became ever more desperate. Then several accidents of good fortune gave the battle to the outnumbered Confederates. Yankee General Howard's brigade came into line at the far Federal right northwest just as new Rebel troops arrived at the right place and time to hit him in the flank. Howard's brigade melted in the firestorm. More Confederate regiments emerged on this crumbling Union flank, and Jeb Stuart's

Battle of Bull Run, July 21, 1861. Union (Gen. McDowell), Confederate (Gen. Beauregard).

cavalry delivered a surprise and devastating charge. The Federal right simply fell apart, spreading panic all along the line. Running for their lives, the Union forces raced back across the bridge and various fords, toward Centreville and eventually Washington.

General Thomas J. "Stonewall" Jackson

McDowell lost 2,900 Yankee soldiers out of about 20,000 engaged. The Confederates lost about 2,000 out of 17,000 engaged. In addition to the men, the North lost artillery and supplies and the morale boost this first battle would have given them — a terrible loss indeed! In contrast, the Southern forces were both jubilant and encouraged.

Our man, J.N. Ruby, had just been promoted to Sergeant Major on July 16, 1861, five days before the blood bath at Bull Run on July 21st. However, his bravery and leadership supported him well and would soon lead to his promotion to 2nd Lieutenant on August 8, 1861. It was this latter promotion that prompted his Co. E, 2nd Wisconsin Regiment to present Ruby with the inscribed M-1862 Colt revolver. Surprisingly, on November 22, 1861, Ruby resigned and was honorably discharged because he wanted to join a cavalry unit; however, he was unable to locate a suitable cavalry opportunity. He returned to civilian life for seven months from November 22, 1861 to June 28, 1862, but again found home life both dull and unrewarding. In frustration, Ruby, looking for fame and fortune, joined the newly formed 20th Wisconsin organized at Camp Randall in June and July 1862. He mustered in August 23rd, going first to St. Louis, then on to Cross Hollow, Arkansas, the enemy falling back at its approach.

On November 4th, the regiment started for Wilson's Creek, joined up with General Totten's command at Ozark on the 11th and reached Wilson's Creek on the 22nd. The primary Union goal in Arkansas was to eventually take Little Rock and thence control of that section of the Mississippi River and Arkansas itself, so that a successful assault on Vicksburg might be launched. In December, Ruby's brigade made a 100-mile forced march in three days to Fayetteville, and was in the savage battle of Prairie Grove where it charged the heights through underbrush and captured a battery of six Rebel cannons. The crossfire of five regiments of the enemy compelled the 20th to retire with a loss of twenty-six killed, 110 wounded and seventy-four missing. Union General Herron wrote Wisconsin Governor Salomen: "I congratulate you and the state on the glorious conduct of the 20th Wisconsin in the great battle of Prairie Grove."

In December 1862, Ruby at long last was offered a promotion to Captain of the newly formed Co. B. 34th Regiment Wisconsin Infantry that had a cavalry unit that J.N. Ruby was chosen to command. Ruby mustered in on December 31, 1862 and left Milwaukee on January 31, 1863.

The brigade arrived at Columbus, Kentucky, February 2nd, where it performed camp and guard duty at Fort Halleck. On May 12th, Cos. B, C, D, F, H and K were sent to Memphis to suppress the Rebels in the area. Ruby's cavalry delivered excellent support service there.

The regiment mustered out at Milwaukee on September 8, 1863. Its original strength was 961 troops — twenty died, 283 were in southern prisons, and 186 were wounded and discharged; only 472 mustered out alive and well.

On March 16, 1864, Captain James Nelson Ruby married Sarah Jane Willoch at Appleton, Wisconsin, Reverend Samuel Fallows presiding. With the help of political connections as a retired officer, Ruby was able to secure a lifelong career with the U.S. Post Office, which grew rapidly following the war. This growth afforded Ruby numerous promotions and advances so that he could provide a prosperous and peaceful life for his family. They had two children — Kate Elizabeth, born February 5, 1868, and James Mason, born November 16, 1873. They resided at 401 Main Street, Oshkosh, Wisconsin their entire married life.

In 1898 Ruby received a $24 per month war related pension for hemorrhoids, insufficient pectoral and aortic heart valves, hypertrophy of the heart, and deformity of the chest wall; also, deafness, liver disease, neuralgia, and general disability.

Captain Ruby died of senility on September 23, 1914 at the ripe old age of 81 years and Sarah died January 12, 1926. They are buried side by side at Riverside Cemetery, Oshkosh, Wisconsin.

References

1. National Archives Register of Enlistments in the U.S. Army 1798-1914.
2. Adjutant General's Official Enlistment Papers, 1st Series 1790-1894.
3. History of the 1st U.S. Cavalry.
4. James N. Ruby Compiled Military Service Record, Federal Pension File and Regimental Histories.
5. Oregon Territorial History 1852-1860.
6. *New York Daily Times*, Saturday, December 1, 1855.
7. The Native Peoples of North America – Northwest Coast and California.
8. The Civil War State By State – Virginia and Arkansas.

Chapter Three

Colonel James Hipkins McNeill The Fighting Preacher of the Confederacy

Confederate armaments identified to a high-ranking officer are always a thrill to find, so Sunday, October 17, 2010, at the Louisville Gun Show (Kentucky) was a lucky day for me.

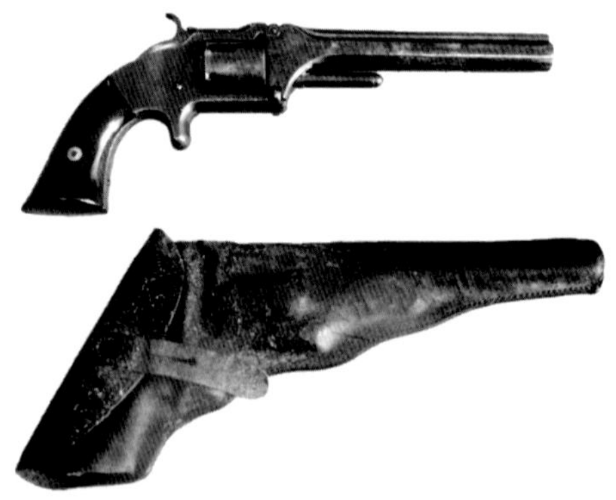

Smith & Wesson Old Army No. 2 Revolver serial number 15740, complete with holster.

Mr. Charles G. (Chuck) Worman, my friend and noted author of numerous gun books, has helped me locate many of the identified or inscribed pistols in my collection. Chuck showed me arms that belonged to Confederate Colonel James H. McNeill, which included his M-1840 heavy cavalry saber and scabbard marked "N. P. Ames, Cabbotville, 1846" and "U.S./N.W.P". Also, included was his Smith & Wesson Old Army #2 revolver, Serial #15740 complete with holster. These weapons descended through McNeill's family and had been owned by only two collectors since they surfaced over twenty-five years earlier at a Nashville estate sale of McNeill heirlooms. Noted collector, Mr. Bruce Jackson, purchased these items

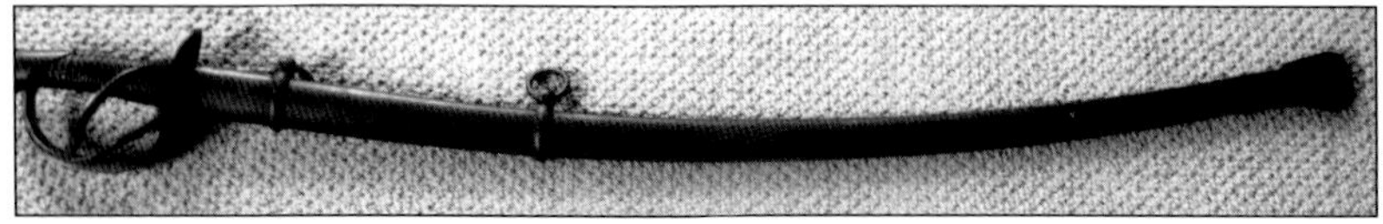

Colonel McNeill's M1840 Cavalry Saber and Scabbard

at the sale and then verified the historical family connection back to James H. McNeill with the help of Marie Douglas, a friend of the family who was conducting the sale. I purchased this grouping because of the immensely interesting life story of Colonel McNeill and for its Confederate connection.

Colonel James Hipkins McNeill at 38 years old.

Very soon after commencing my research on McNeill, I was blessed to make contact with a living relative, Thomas McNeill Rose III, residing in South Carolina. Mr. McNeill Rose was an active historian, very familiar with the Civil War and his own family history. He was extremely helpful, providing me with a superb amount of information including a photograph of the very handsome and dashing Colonel McNeill.

It's a rare Civil War soldier who was educated at Princeton Theological Seminary, Yale University, The University of North Carolina, Union Theological Seminary, and Delaware College. In addition, he became an ordained Presbyterian minister in May 1849, and founded a Presbyterian Church in Pittsboro, North Carolina. Then, he served as a chaplain in the Confederate Army from August 1861 to September 1862. He then enlisted as a captain of Co. A, 5th North Carolina Cavalry (63rd State Troops). On October 26, 1862, he was promoted to major and then, due to his zealous fighting spirit and intelligence, he was promoted to colonel on November 24, 1864.

James McNeill was born May 25, 1825, in Fayetteville, Cumberland County, North Carolina, the son of George and Minerva McNeill. The family was prosperous and lived well, being able to send all three of their children to college. George, the youngest son, graduated from Princeton and Jane, the oldest child, graduated from the University of North Carolina at Chapel Hill.

Letters between Jane and James during his college years show evidence that James was a serious and talented scholar. While at Princeton, he met and courted Kate Chamberlain, daughter of Dr. Palmer and Grace Chamberlain. They became madly in love and were married October 16th, 1848 at Wilmington, Delaware. Kate and James had five children:

1. George Palmer, born September 16, 1849; 2. James Hipkins II, died in infancy; 3. Kate Ruffin, born July 19, 1854; 4. Charles Joseph, died in infancy; and, 5. John Rose, born August 9, 1861.

In 1848, shortly after their marriage, James founded a Presbyterian church in Pittsboro, Chatham County, North Carolina, which he nurtured and expanded until 1853. He was then appointed to the American Bible Society in New York City and moved his family to Elizabeth, New Jersey. After several years, James was elevated to General Secretary of the Bible Society. He was a noted linguist and did much research in Greek history and language. During these years, he translated the Bible into the Spanish language, an unbelievable accomplishment.

Upon the outbreak of the Civil War, James moved his family back to Fayetteville, North Carolina, and volunteered as a chaplain in the Confederate Army. During this year, James became more and more convinced that he needed to become active in the army. So, on September 30, 1862, he mustered into Company A, 5th North Carolina Cavalry as a captain. Because of his excellent character and education, he was promoted to major on October 16, 1862.

The Fifth Cavalry, Sixty-Third Regiment, North Carolina Troops, was organized at Garysburg, North Carolina in the early fall of 1862. It was composed of companies enlisted under the Partisan Ranger Act of the Confederate Congress. The act promised many special rights and privileges to the Rangers, not the least of which was the clause saying that all property captured from the enemy became at once the private property of the captor. Most of the officers and many of the men had previously been in the army; thinking that the Ranger Service would afford greater scope for individual exploit and consequent glory, they had managed by one means or another to get a transfer.

The Staff Officers were:

Peter G. Evans, Colonel, Chatham County
S.B. Evans, Lieutenant Colonel, Goldsboro
James H. McNeill, Major, Fayetteville, North Carolina.
J. Turner Morehead, Adjutant, Greensboro, North Carolina
George Haigh, Sergeant-Major

Colonel McNeill was a truly outstanding officer. General Barringer wrote of him in his nomination for promotion, "James McNeill has risen to the highest standards of gallantry and efficiency. He is an outstanding citizen of North Carolina, a gentleman of education and attainment of high personal character — great energy of purpose — true courage and strongly dedicated to this service."

McNeill was known for two admirable characteristics. War conditions permitting, on Sunday mornings he would put on his vestments and conduct church services for the men who were eager for moral support and spiritual underpinnings. He was exceedingly brave, often leading cavalry charges rather than hanging back. He never asked his men to do anything he wouldn't do. The 5th North Carolina Cavalry saw a tremendous amount of hot fighting action in thirty-four battles, often times in hand-to-hand combat as follows:

New Bern (March 14, 1862), White Hall (December 16, 1862), White Oak River (Co. E.) (March 7, 1863), Brandy Station (June 9, 1863), Middleburg (June 17, 1863), Upper Ville (June 21, 1863), Gettysburg (not engaged) (July 1-3, 1863), Fairfield (July 3, 1863), Hagerstown (July 5, 1863), Jack's Shop (September 22, 1863), Russell's Ford & James City (October 10, 1863), Culpeper Court House (October 11, 1863), Auburn Mills (October 14, 1863), Manassas Junction (October 15, 1863), Buckland (October 19, 1863), Parker's Store (November 29, 1863), White Hall (May 6, 1864), Spotsylvania Court House (May 8-21, 1864), Goodall's Tavern (May 11, 1864), Ground Squirrel Church (May 11, 1864), Wilson's Wharf (May 24, 1864), Hanovertown (May 27, 1864), Haw's Shop (May 30, 1864), Hanover Court House (May 31, 1864), Ashland (June 1, 1864), Riddell's

Members of the Sixty-Third Regiment beginning top left: Colonel Peter G. Evans, Colonel James H. McNeill, Major John M. Gallaway, Adjutant and 1st Lieutenant J. Turner Morehead, 2nd Lieutenant (Co. A) James D. Nott, Private (Co. A) James Kirkpatrick, and Private (Co. A) T.F.R. Rose.

Shop (June 13, 1864), Point of Rocks (June 1864), Malvern Hill (June 1864), Herring Creek (June 1864), Petersburg Siege (June 1864-April 1865), Five Forks (April 1, 1865), Namozine Church (April 3, 1865), and Appomattox Court House (April 9, 1865).

Two battles were of special significance to Colonel McNeill — Middleburg and Five Forks.

Middleburg, Virginia (from *North Carolina Troops, 1861-1865*) – Confederates had pretty good luck at the collision that occurred at Middleburg, June 17, 1863. The Fifth Squadron of the Sixty-Third was ordered at about dusk to charge on some Yanks in the road. However, these men were simply decoys. A whole Federal regiment was dismounted and hiding behind a stone fence. The hidden regiment opened fire, killing one man and disabling about twenty men and horses. Only seventeen men of the squadron followed the captain through the fire. Of the remainder some tried to turn back, some dismounted and took shelter behind the stone fence. Fortunately, the rear of the 63rd Regiment was close at hand, dismounted, got over the fence and attacked the enemy in flank. Still more fortunately, a Virginia regiment was passing on a road perpendicular to the road charged upon. This regiment heard the firing, halted and was ready to receive the enemy as they gave way before the flanking action. Nearly the entire regiment of Federals was captured,

Battle of Middleburg, Virginia, June 17, 1863.

about 800 men. The 63rd lost two men killed, about twenty wounded, among them were three lieutenants. The greatest loss was Major McNeill, shot straight through the hip. McNeill was shot while handling his portion of the flank attack very skillfully. He was moved to Military Hospital #4 in Richmond, Virginia, where he recuperated in about two months.

Five Forks, Virginia (Chamberlain's Run) according to Confederate reports this was the most fearful and fiercest battle the 5th North Carolina Cavalry had ever seen. Because of its proximity to Richmond, the capital of the Confederacy, a Union report called it, "The best cavalry fight ever." The question is why was Colonel McNeill repulsed at the ford the morning of March 31, 1865, and what part did Colonel McNeill and his grand regiment play in this awful tragedy.

At the same time that Colonel Cheek received his orders for the action, Colonel Gaines, of the Nineteenth and Colonel McNeill received theirs from General Rufus Barringer. Colonel Cheek executed his magnificently, and so did Gaines and McNeill. The Sixty-Third was in front. A small detail was sent, mounted, to the right as scouts under Captain S. A. Grier. McNeill and Gaines were told to dismount their regiments, go to the ford, and cross in columns of fours. The Sixty-Third was to deploy in line of battle to the right of and below the ford; the Nineteenth to follow and deploy to the left and above the ford, completing and connecting the line between Cheek and McNeill so they could drive

Battle of Five Forks (Chamberlains Run), Virginia, March 31, 1865.

Brigadier General Rufus Barringer

the enemy. That was the plan and the orders.

The road crossed that ford at right angles. The water there, one hundred and fifty yards below Cheek, was over the men's waists. Of course, it was much deeper for McNeill's group than where the Nineteenth was — so deep, so impassable by reason of briars and swamp undergrowth and other obstructions of fallen timber on both sides of the stream, that it could not be crossed for battle except at the ford. Men were shot down in the ford, swept off by the current and actually drowned before their comrades could pull them out. This was the unfortunate place the Sixty-Third and Nineteenth had to cross under their orders.

Across the stream, from the road up to Cheek's right, was a body of small and large timber extending forward almost to the enemy's entrenchments; immediately to the right of the road was open ground, sparsely wooded, thirty-five to fifty yards wide up to the enemy's works. From the stream the ground rose rapidly to the enemy's lines and works, which were about 200 yards from the stream with their left point being almost opposite to what was to be McNeill's right. McNeill's intended right, across the creek, would overlap their left slightly. From the place where they dismounted, the Sixty-Third and Nineteenth moved rapidly towards the ford, a large body of Sheridan's Federal Cavalry was on the Confederate side of the stream, a fierce fight ensued and they were driven pell-mell across the ford back to their works.

In this affair Colonel Gaines lost his right arm. Grandly and gloriously, with Colonel McNeill in the lead, our regiment crossed that ford under a galling, withering fire from Henry rifles, that shot sixteen times without reloading, fired by an entrenched enemy. As they crossed, McNeill gave the proper orders loud and clear—each company captain, as his company landed, repeated the order and quickly the regiment was in line of battle to the right of the road waiting for the Nineteenth to cross.

General Barringer sent a courier to learn of McNeill's situation and progress. McNeill was advancing slowly, waiting for the Nineteenth to form on his left, and the fire was so furious that it was better to advance than to stand. To the courier Colonel McNeill coolly said, "Please tell General Barringer I am all right and advancing slowly, waiting for the Nineteenth to form on my left so we may charge and carry the day. Ask him, please to hasten the Nineteenth over."

Colonel James McNeill's gravestone at Crosscreek Cemetery, Fayetteville, North Carolina

Unfortunately, someone *blundered!* To distract the fire on the two Confederate regiments, W.H.F. Lee ordered a Virginia regiment to charge across the ford mounted, just when the Nineteenth was steadily marching forward to cross. Misunderstanding their orders, only a squadron of the Virginians rushed over and up the incline of the road. The Nineteenth closed right in behind them. About two companies of the Nineteenth crossed behind that squadron. Inexplicably, that squadron broke and fled back across the narrow ford; panic stricken, they were breaking the line of the Nineteenth and pushing them down into the deep water at the lower side of the ford. The enemy was so exultant over their sight of the fleeing squadron that they advanced and redoubled their already furious fire on McNeill and the ford, where the column of the Nineteenth was now helplessly cut in two by that charging mass of mounted men.

While McNeill's ammunition was almost exhausted, D.B. Coletrane, standing near him said, "Colonel, I have only two cartridges; shall I use them or hold them?"

"Keep them, you may need them more in a moment," coolly and calmly answered McNeill in the face of well-recognized and terrifying danger. And there he and his regiment were, fighting and firing their last shots.

Captain C. W. Pearson was talking to Colonel McNeill when he saw a man a little to the right, run from a large pine to another pine. Very soon a puff of smoke came from the tree. Colonel McNeill was shot in the head, killed instantly. Some of the men picked him up but only got a short distance when the entire line gave way. The loving men carried off their leader as they fired their last shots into the swarming ranks of an overwhelming, advancing foe. Sadly, McNeill was killed only ten days before General Lee's surrender at Appomattox Court House. He left a wife and three children. A full military funeral honored McNeill as he was buried in Crosscreek Cemetery in Fayetteville, North Carolina.

Chapter Four

THE COOK FAMILY PROVIDES FOUR SOLDIERS TO FIGHT IN THE "GREAT WAR OF THE AMERICAN REBELLION"

A matched pair of Smith & Wesson No. 2 old Army revolvers complete with holsters, .32 Cal. 6-inch barrels, Serial #s 7837 and 12905. This pair was carried by Second Lieutenant Samuel Cook, Co. H, 13th Indiana Volunteer Cavalry.

In March 2004, I purchased three Smith & Wesson #2 Old Armies from Dave Taylor, a very respected Civil War dealer from Sylvania, Ohio. Dave had obtained these three revolvers, two matching holsters, a shooter's box, and Civil War documents directly from Cook family descendents. Accompanying the guns is the typed family record detailing the history on the revolvers and holsters, as they understood it.

The senior family member, 36-year-old Second Lieutenant Samuel Cook, enlisted February 6, 1864, for three years at New Albany, Indiana. He was assigned to Co. H, Thirteenth Indiana Volunteer Calvary, 131st Regiment under Captain W. H. Clendening. Samuel Cook's youngest son, Private George W. Cook, enlisted concurrently with his father in the 145th Indiana Volunteer Infantry as a drummer boy. Remarkably, Private George Cook was only eleven years and eight months old on his enlistment day!

A brief summary of the exciting action of the 13th Indiana Cavalry follows:

- By rail to Nashville, April 30, 1864.
- Moved to Huntsville, Alabama, on garrison duty until November 1864.
- Repulsed Buford's attack on Huntsville, September 30-October 1, 1864.
- Siege of Murfreesboro, December 5-12, 1864.
- Siege of Decatur, Alabama.
- Battle of Nashville, December 15-16, 1864.
- Battle of Hillsboro, December 29, 1864.
- Battle of Leighton, December 30, 1864.
- Siege of Mobile, March 28-April 12, 1865.
- Raid through Alabama, Georgia, and Vicksburg, Mississippi, April 17-May 22, 1865.
- Mustered out at Vicksburg, November 18, 1865.
- Total dead of wounds and disease — 142 soldiers.

Lieutenant Cook contracted erythema, an affliction of the legs and feet, in April 1865 as a consequence of extreme exposure to inclement weather and continuous hard riding while on a raid from Mobile to Montgomery, Alabama. Fortunately, he recovered in thirty days and was able to rejoin his unit.

OLIVER P. MORTON,
GOVERNOR OF THE STATE OF
INDIANA,

TO ALL WHO SHALL SEE THESE PRESENTS, GREETING.

Know Ye, That reposing special confidence in the patriotism, valor, fidelity and abilities of Samuel Cook I have appointed, and by virtue of the authority vested in me as Governor of the State of Indiana, hereby commission him Second Lieutenant in the 13th Cavalry One Hundred Thirty First Regiment INDIANA VOLUNTEERS raised under the authority of the President of the United States, and the Laws of Congress, to serve during the period for which said Regiment was called into the service of the United States vice Wm. M. Munson promoted and he will be GOVERNED, OBEYED AND RESPECTED according to the Rules and Articles governing the Volunteer Armies of the United States.

Given under my hand at Indianapolis, in the State of Indiana this Twenty Second day of March One Thousand Eight Hundred and Sixty Five

[illegible] — Adjutant General of Indiana.

O. P. Morton — Governor of Indiana.

Document commissioning Samuel Cook as Second Lieutenant, 13th Cavalry, 131st Regiment, Indiana Volunteers signed by Oliver P. Morton, Governor of Indiana.

Before and after the Civil War, Samuel Cook was a revered blacksmith in his hometown of Mitchell, Indiana. However, by the time he was 52 years old, in 1880, Samuel was about fifty percent disabled for manual labor because his feet would swell up more and more and become painful as the day went by. Samuel had to have special boots made to allow for the daily swelling. By 1886, Samuel Cook could only perform manual labor about twenty-five percent of the time and his pension was increased to $24 per month.

Samuel died of pneumonia on March 13, 1895 at 67 years of age, leaving his wife, Agnes, in destitute financial condition. On her "Widow's Application for Accrued Pension," an affiant stated that, "Claimant, Agnes Cook, has no property except a house and lot in Mitchell, Indiana worth about four or five hundred dollars. That she is without other means of support other than her own daily labor and is dependent on others not legally bound for her support, that she has no income now nor hasn't had since the death of her husband except contributions of others."

Drummer Boy, Private George W. Cook, suffered no injury during his twenty months of service, but he did contract diseases that resulted in stomach gastritis, causing dilation of the heart, torpid liver, piles, and an attack of appendicitis. These problems were enumerated and explained in his application for a pension on March 21, 1911 when George was 59 years of age. At this time, George W. Cook was operating his own mining company in Leadville, Colorado.

In reviewing the details of his pension application, I discovered George Cook was the Mayor of Leadville, Colorado, in 1886. At this time Leadville was a rip-roaring mining town, high in the Colorado Rockies, so Mr. Cook must have led quite an exciting life. George married his first wife, Jennie Godman, on September 6, 1873, with whom he had one daughter, Georgie. Jennie died January 6, 1887 of severe fever. George later married Nina F. Purdy in 1891, with whom he had a son, George E. Cook, born in Denver, Colorado, May 11, 1893. George encountered many business problems in the mining industry, which put him under great mental stress. Unfortunately, George was declared mentally incompetent November 14, 1913, and died in an insane asylum on December 17, 1916.

Lieutenant Cook's oldest son, Private John A. Cook, 13th Indiana Volunteer Cavalry, was killed in action near Montgomery, Alabama, and was buried in the cemetery closest to the capital of Alabama. John A. Cook was only 16 years old at his enlistment and was a bugler for his unit. He was only 18 years old when he died on June 6, 1865. A letter returning his effects to his family reads as follows:

"1 coat; 2 pair of pants; pair of drawers; 1 shirt; 1 pair of socks; 1 rubber blanket; 1 bed blanket; 1 canteen, haversack; bugle; $1.55 in money."

His death was caused by a gunshot wound to his foot and camp diarrhea. The following is an interesting letter sent home by John A. Cook:

> "Huntsville, Alabama
> June 19, 1864
>
> Dear Brother, Mother and Sister:
>
> I take the pleasure of writing to you to let you know that I am well and enjoying good health. All of the soldiers is leaving but 3 or 4 regiments. We have a nice time here. We are going to move our camp in a few days up by the fort. Our captain is Provost Marshall and Father is a going to town to draw rations for eighteen men and wile stay there as long as the regiment stays here. All of the boys from Mitchell is well and we got that box of provisions and everything was spoilt but the peaches and butter. Frank Hamiltons cake was good and Edward Eversole got his box the same time that we did and some person stold a ham of meat and some tobacco. The next time you send anything let it get cold and don't put very much butter init for the butter gets strong and spoils the cake. Father wants you to get some Hamers whiskey and send it in the next box that they send either a gallon or a half a gallon. We can get whiskey down here but it ant no count. The Fifth Ohio can drink like the Thirteenth, for just as regular as the day comes, they come over to our sutler and commence swearing that they can clean us out. There is three fellows in Capt Hermans Company that had four

fights with them, and the old Thirteenth come out victorious and as it happens today they haven't come over yet. Every time any of them comes over they get whiped. There is a whole lot of Missouri regs that come in town the other day and they called us conscripts and it wasent but a very few hours some of them come over and they got whiped like the thunder and we have got them a fraid of us, and here is the Fourth Minnesota that thought that they could run over us and one of them got whiped last night. The only regs that gets along together is the 59th Indiana Infantry and the old bloody 13th Indiana Calvary. Sprint Mannington is inspector of the block house along where they have them at bridges to fight in to keep the rebels from burning the bridges. The next time you send a letter I want you to send me two dollars. I don't need money, but I don't want to get out. I can get two dollars for one at pay day and a good many of the boys lends it that away.

Write soon
John A. Cook
Co H 13th Ind. Cav.
Huntsville, Alabama
Give my respects to all
of the boys and also the girls."

Of course, the family never recovered from the loss of two of their sons.

The fourth family member to serve in the Civil War was Corporal Jacob Deisher, son-in-law of Lieutenant Samuel Cook, who was married to Sarah Cook, Samuel's only daughter. Corporal Deisher was assigned to Co. D, 155th Regiment of Illinois Infantry from February 28, 1865 until his Honorable Discharge September 4, 1865 at Murfreesboro, Tennessee.

During his service he contracted chronic diarrhea and neuralgia of the face and head from general exposure. For ten days he had no blanket, and spent much of the time in mud and water near Camp Tullahoma, Tennessee. In 1884, he claimed he was fifty percent disabled for manual labor and his left eye was greatly impaired by neuralgia. The 155th Illinois Infantry participated in the Battle of Nashville and the Battle of Chattanooga. They also defended blockhouses on the Nashville and Chattanooga railroad until September 1865.

After the war, Jacob Deisher worked on railroads until 1884, when his disabilities from the war caused him to be unfit for manual labor required by the railroads. Jacob then became a clerk and bartender in a saloon. Jacob applied for a pension on May 29, 1884, which was granted October 15, 1885 in the amount of $6 per month, which was raised to $17 per month by 1898 when he lost the vision in his left eye.

Sarah and Jacob had one child, a daughter named Lydia, born March 5, 1875. This was fortunate because the Deishers were never well off financially and could hardly afford more children. Jacob Deisher died on October 3, 1906, from angina pectoris. Jacob's widow, Sarah, applied for and received a pension of $30 per month. She lived frugally with boarders in her home until her death April 21, 1928.

Cook family records indicate the pair of Smith & Wesson's with matching holsters were carried by Second Lieutenant Samuel Cook and the third Smith & Wesson Old Army #2 in the shooters box was carried by Corporal Jacob Deisher.

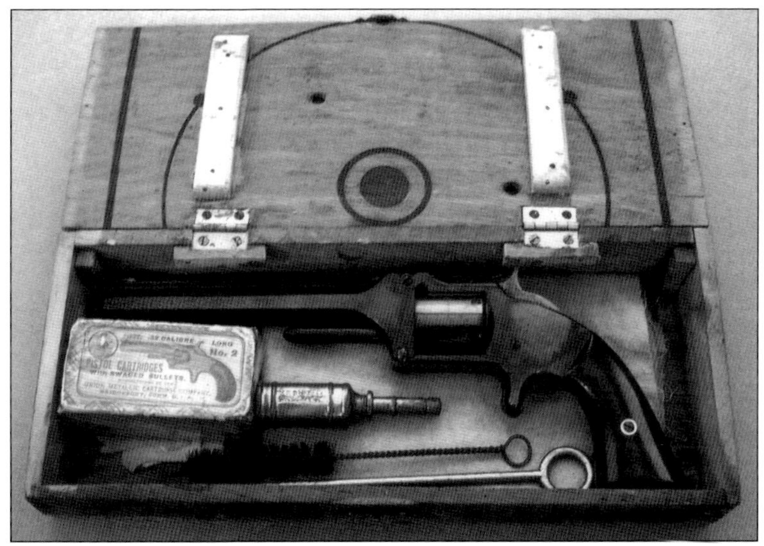

Smith & Wesson old Army No. 2 in a shooters box, .32 Cal., 6" barrel, Serial #29500. This gun was carried in the Civil War by Corporal Jacob Deisher, son-in-law of Lt. Samuel Cook, who was married to Sarah Cook, Samuel's only daughter. Corporal Deisher served in Co. D, 155th Illinois Infantry.

Chapter Five

Four Very Rare Inscribed or Identified Civil War Revolvers

For those of us who enjoy legitimately inscribed or identified Civil War arms and their stories, it is natural that we gravitate to the famous name revolvers supplied in large numbers by Colt, Remington, and Whitney, etc.

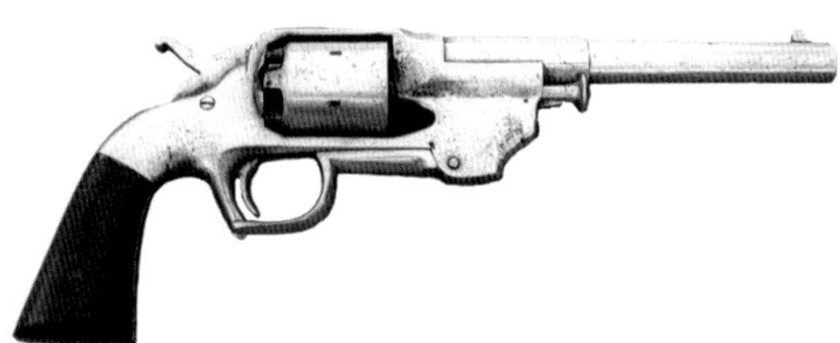

Fig. 1: Allen & Wheelock, .44 Cal. Army, 7.5-inch barrel, Serial #88, issued to Private John E. Cranston, 3rd Michigan Volunteer Cavalry.

But for the serious Civil War aficionado, it would seem that this interest would also include the very seldom seen, rare, or concealed side arms that saved many a soldier's life, when engaged in fighting at close quarters rather than across the battlefield with cannon and rifles.

The purpose of this chapter is to illustrate four unusual identified arms and the interesting details of their service records and owners.

The first is the Allen & Wheelock Percussion .44 Caliber Army Revolver (Fig. 1). These big old warhorses are very elusive. Of the estimated 750 to 1,000 revolvers manufactured, only 198 are known to have been purchased by the U.S. government and issued to Civil War soldiers.

On December 31, 1861, the U.S. Army Ordinance Department bought 198 Allen & Wheelock .44 Caliber, Center Hammer Revolvers from William Read and Sons of Boston for $22 each. It would seem that survival rates are low, especially for those few that can be documented as having been issued to a specific soldier. In terms of relative rarity, these guns are rarer than any of the Colt Dragoons, the Confederate Leech & Rigdon, or even the Griswold, the Spiller & Burr, etc.

The one illustrated is a first model, Serial #88, 7-1/2" barrel. An unusual feature is that the trigger guard actuates the rammer that seats the ball securely in the cylinder. Fortunately, through records available from the Springfield Research Service and National Archives, it is possible to verify by serial number this revolver's issue to Pvt. John E. Cranston of the 3rd Michigan Cavalry, Co. I. The 1995 edition of the *Springfield Research Service* on serial numbers, Volume Four, lists the usage of fourteen (14) Allen & Wheelock Army Revolvers, nine (9) by Co. I, 3rd Michigan Volunteer Cavalry, and five (5) by Co. A, 2nd Michigan Volunteer Cavalry, commanded by Captain George Armstrong Custer.

Private Cranston was born in 1840 in Tyrone Township, Livingston County, Michigan. The Cranston family members were Christian Methodists and one of the most prominent families in Tyrone, having descended from John Cranston, one of the early governors of Rhode Island. John Cranston's father, Samuel, was a notable soldier of the American Revolution.

Private Cranston enlisted February 16, 1864, at Pontiac, Michigan. He mustered out February 12, 1866, at San Antonio, Texas, and was discharged at Jackson, Michigan. February through May 18, 1864, the 3rd Michigan guarded St. Louis, Missouri. They proceeded to Little Rock, Arkansas, reporting to Major General Steele and assisted in driving Confederate General Shelby beyond the Arkansas River. The regiment then moved by steamship to New Orleans, Louisiana, and were engaged in the Battle for Mobile, Alabama, in April 1865. After the fall of Mobile, they were employed on outpost duty in Alabama, Louisiana, and Texas until the surrender of Confederate forces east of the Mississippi River. In all, the 3rd Michigan Cavalry was engaged in battles or skirmishes twenty-four times.

John Cranston had two brothers and three cousins who served in the Civil War. John was a successful farmer before and after the war, fathered six children with his wife, Elizabeth, and unfortunately died at forty-five years of age in 1885 from a logging accident.

The second revolver, a Belgian-made 7.65 MM (.30 Cal.), Pinfire (Fig. 2 and Fig. 3) with a folding trigger is seldom seen, especially

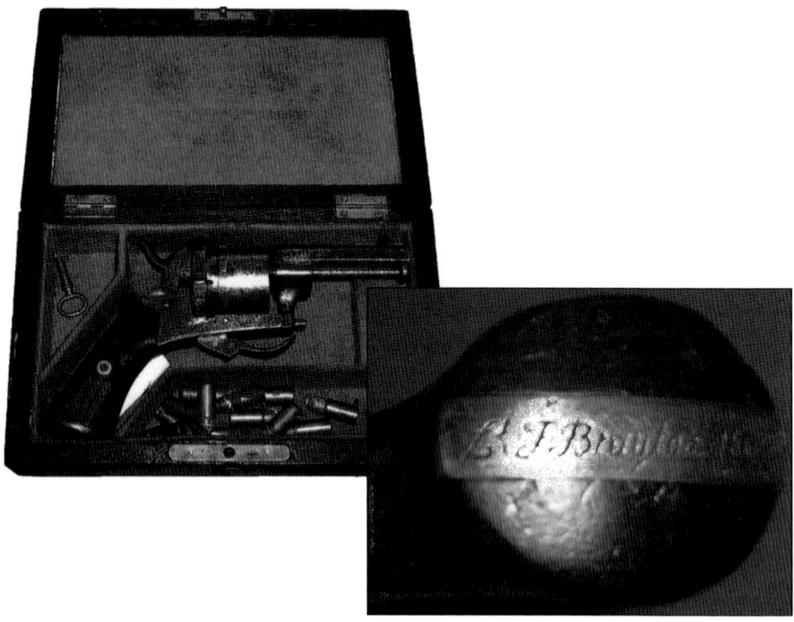

Fig. 2: Belgian made 7.65 mm (.30 Cal.), pinfire, cased with metallic cartridges. Fig. 3 at right: The inscription reads - LT. T. Brayton, 1862.

with a Civil War inscription, since these were relatively inexpensive guns. The great advantage to this gun is that it shot self-contained cartridges and was easily concealed. The butt is inscribed, "Lt. T. Brayton – 1862". First Lieutenant William Thomas Brayton was the Chief Bugler, 3rd Wisconsin Cavalry, Co. C and Co. H. He was born in Kingsbury, New York, in 1833 and enlisted in Co. C as a bugler, October 19, 1861 at Janesville, Wisconsin, then Co. H, November 19, 1862, at Milwaukee, Wisconsin. He mustered out September 24, 1865, at Fort Leavenworth, Kansas.

Before and after the war, Brayton was a music teacher and musician, which gave him special qualifications for the critically important job of Chief Bugler. Most military commands were transmitted to the engaged-fighting troops by bugle so the success or failure of the cavalry bugler could often determine the outcome of a skirmish. The 3rd Wisconsin Cavalry participated in sixteen battles including Fort Gibson, Perrysville, Jackson Port, and Lexington. Because of his musical talent, Lieutenant Brayton also played in the Regimental Band at Fort Leavenworth, Kansas. He passed away peacefully in his sleep on March 22, 1895, at sixty-two years of age.

Another very unusual Civil War sidearm is this Allen and Thurber Pepperbox (Fig. 4 and Fig. 5) inscribed, "Jacob Muhl", which was carried by Corporal Muhl throughout his service in four companies – Co. E, 40th N.Y. and Co. F, 55th, New York, Volunteer Infantry and then Co. E, 13th New Hampshire and Co. B., 2nd New Hampshire Volunteer Infantry. In February 1865, Corporal Muhl was part of an elite division of sharpshooters owing to his excellent marksmanship, which he had worked hard to attain during his four-plus years of Civil War service. The pearl handled Sheffield Knife, which came with the gun, is from the Muhl family; together, they comprise an uncommon duo representing our Civil War.

Jacob Muhl is an example of one of the many brave immigrants who served in America's Civil War. He was born in Ruppenheim, Germany, in 1839, and immigrated to America in 1858, settling on the outskirts of New York City where he had acquired a small farm. In September 1861, he enlisted for a three-year term and was not finally discharged until December 1865.

During his service with the 55th New York, Corporal Muhl was involved with the defense of Washington, D.C., and went through seven battles including Manassas, the siege of Yorktown,

Fig. 4 & 5: Allen Thurber & Co. Pepperbox, 6-shot, double action, .30 Cal., 3.25-inch barrel. The pearl handled Sheffield knife came with the gun. Inscription on the barrel reads, "Jacob Muhl". Corporal Muhl served in four Civil War units: Co. E, 40th New York; Co. F, 55th New York; Co. E, 13th New Hampshire; and, Co. B, 2nd New Hampshire. He was part of an elite division of sharpshooters.

the battle of Williamsburg, and the seven-days battle at Harrison Landing, Maryland. After Harrison Landing, Jacob Muhl fell ill with rheumatism, lumbago, and dysentery, with hemorrhoids and lung disease. He was sent to Hammond General Hospital at Point Lookout, Maryland on July 20, 1862, and was honorably discharged on December 24, 1862.

Having recovered, on August 11, 1863, he re-enlisted with Co. 13, New Hampshire Volunteer Infantry. During this term, Muhl fought the siege of Richmond and Petersburg, June 16, 1864, until April 2, 1865, plus six other battles including Cold Harbor, Fort Harrison, and the Battle of Fair Oaks. The 13th New Hampshire was the first Union Military Regiment to carry their colors into the Confederate capital of Richmond, Virginia, on April 2, 1865. Muhl's Regiment lost five officers and 175 enlisted men.

Following the war, Jacob Muhl became a wholesaler of liquor, wine, and ale. His business prospered and on September 15, 1867, he married Elizabeth Weinberger, also a German immigrant, and together they had four children — Jacob, Wilhemina, Josephina, and Katherine. Jacob and Elizabeth became well-to-do New Yorkers and purchased four homes, the last of which was located at 157 West 61st Street, New York, New York, and was said to be very elegant.

Jacob Muhl died February 1, 1906, at sixty-seven years of age of pneumonia, and was buried in Evergreen Cemetery, New York. His wife, Elizabeth, died only months later in November 1906. Her death was attributed to a broken heart.

This last example of interesting, non-official Civil War side arms is a Smith and Wesson Model 1, Second Issue, 22 Caliber Revolver with, "William H. Herrick" inscribed on the backstrap of the revolver (Fig. 6 and Fig. 7). The barrel has been shortened 1" to 2-1/4" and the ejector rod has been discarded in order to make an easily concealed hide-away gun. The gun came from the Herrick family with a gilt-framed tintype picture of William H. Herrick (Fig. 8). In the picture he is well dressed in the style of a gentleman. Immediately after his discharge from the war, Herrick went to Ann Arbor, Michigan Medical School, and in 1866 graduated as a family physician.

Private William H. Herrick mustered into Co. B, 87th Ohio Volunteer Infantry on June 10, 1862, at Camp Chase, Ohio (Co-

Fig. 6, 7, & 8: Private William H. Herrick and his Smith & Wesson Model No. 1, 2nd issue, .22 Cal. Revolver, barrel sawed off to 2.25" and ejector removed to make a hideaway gun. The gilt framed tintype of the owner is faded. Backstrap inscription reads "William H. Herrick". Pvt. Herrick served Co. B, 87th Ohio Volunteer Infantry and became a prisoner of war in Richmond, Virginia's prison.

lumbus). On June 12th, the regiment left to defend Baltimore, Maryland, until July 28th and were attached to a railroad brigade, 8th Army Corps, and ordered to Harper's Ferry, West Virginia. After numerous skirmishes, the regiment was sent to the defense of Harper's Ferry, September 12th to the 15th, where they were defeated by the Confederates and the entire regiment was confined to the prison in Richmond, Virginia. Herrick was paroled from prison in September 1862, and mustered out at Camp Chase, Columbus, Ohio. Herrick's wartime service was only four months long, but certainly action-packed and eventful.

Dr. William Herrick married Emeroy Travis, August 8, 1866, in Cleveland, Ohio, and proceeded to have four children — Harry, Nina, Maud, and Mary. Dr. Herrick prospered and practiced family medicine until his death on May 29, 1920, at the ripe old age of seventy-six.

The four guns described above are all very different and unique, but illustrate the fact that endless types of weapons served as extra protection during America's Great Rebellion.

Chapter Six

Sergeant Al D. Fobes of the Famous 11th Indiana Volunteers From Affluence to Prison

Albert D. Fobes was born August 12, 1842, in the city of Lima, La-Grange County, Indiana. He was fortunate to be born to rather affluent parents, Joe and Anna Fobes, who were the proud owners of an 800-acre farm. Indiana was known as part of the great breadbasket of America and contained some of the richest and most productive farmland in our country.

Typical Zouave uniforms

All his life Al had played military games and dreamed of the day he might become a real soldier engaged in an important war. He could picture winning glory and medals for heroic deeds performed on the field of battle. When Al was seventeen years old he joined the Montgomery Guards at Crawfordsville, Indiana, under the command of Captain Lew Wallace. The Montgomery Guards were reported by the *Crawfordsville Review*, "...to be the best military company in the state. It will be recollected that this company, at the time it was here during the military encampment last July, was considered one of the best drilled and best uniformed of those present. Since that time they have much improved in military tactics and can, we are told, execute six hundred and thirty evolutions by the tap of the drum. We have no doubt but that the company will always maintain its high position so long as it keeps Captain Wallace at its head."

Al Fobes was fortunate to make the acquaintance of Lew Wallace, who would become a legendary man during his lifetime. At only thirty-four years of age Wallace was promoted to Major General, which made him the youngest officer to hold that rank. In addition, after the Civil War, Wallace wrote a number of books including *Ben Hur*, making him one of America's most celebrated authors.

Captain Wallace was seeking to improve his Montgomery Guards and before long, Wallace came across a magazine article about France's Algerian Zouaves. Their colorful uniforms and dashing drill appealed irresistibly to Wallace's sense of the theatrical, and he had no trouble converting his company to the Zouave System. It took them awhile, through benefit balls, parades, and exhibitions to raise the money, but eventually they acquired proper Zouave uniforms—Kepis with red cloth hanging behind, red and blue Greek tunics, baggy gray breeches, and gaiters. Thus costumed, the clerks, mechanics, farm boys, and Wabash College students looked like Veteran Legionnaires. Wallace persisted and kept the company alive until war broke out in 1861. At this time all the members of the Montgomery Guard, including Al Fobes under newly appointed Colonel Wallace, enlisted in the Union Army.

Indiana's Governor Morton had promised Wallace his appointment to Colonel, as well as leadership of one of the regiments in return for raising six regiments for the State of Indiana. Wallace suggested numbering the new regiments Sixth through Eleventh and chose command of the last.

The Eleventh Indiana followed the Zouave System of warfare exclusively. It was adaptable to a sort of commando warfare, in which men would take cover instead of marching in ranks to be slaughtered, in which they would fight on their belly as well as on foot. Double time was habitual, and commands were given by bugle rather than by voice, so that a commander could control his regiment when it was stretched out of sight in a forest or over a mountain.

Private Al Fobes would soon be leaving for unknown battlefields. His proud and appreciative parents presented him with a Colt M1849 pocket pistol inscribed on the back strap "Al D. Fobes — Company E, 11th Indiana." Within a few weeks, the Eleventh paraded through a continuous ovation to the Indiana State House, where they received their regimental colors, hand sewn by the patriotic ladies of Indianapolis.

After the presentation of the colors, Wallace gave an impassioned speech reminding his regiment of the Battle of Buena Vista in the Mexican War, in which the Second Indiana was wrongly discredited. Now, insisted Wallace, was the chance to avenge this wrongdoing. He then had his men kneel, raise their right hands, and take the oath to "Remember Buena Vista!" Thus was established the regimental motto.

The audience went wild with applause, and the motto was adopted by the rest of the Indiana regiments. The event was reported in newspapers around the country, and was illustrated with a full page engraving in *Harper's Weekly*.

The Eleventh Indiana went by train to Maryland, with tumultuous receptions en route in Indianapolis and Cincinnati. When the train stopped at Grafton, where the railroad forked, Wallace discussed the situation with Brigadier General Thomas A. Morris. Cumberland was halfway between Grafton and Hagerstown, making it too

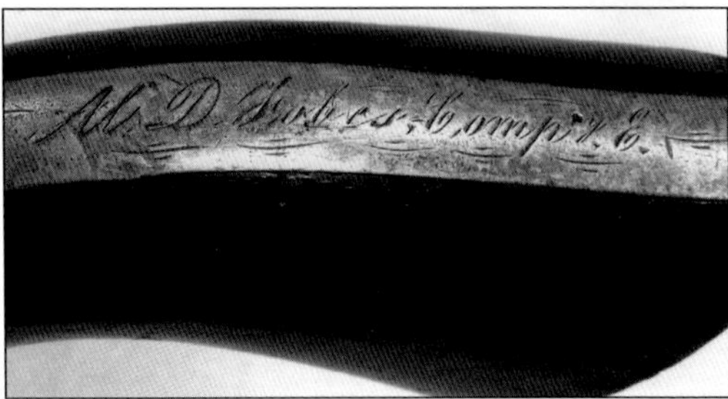

The inscription reads, "Sargeant Al D. Fobes, Comp. E – M1849 Colt Pocket Pistol

far for support from either General Morris or General Patterson, and hence vulnerable to attack from a Confederate Garrison in Romney, some forty-five miles to the southwest.

Wallace decided not to wait for trouble, but to take the initiative and attack Romney before going into quarters at Cumberland. To deceive the Confederates, who doubtless had informants in the city, Wallace went first to Cumberland, examined the possible campsites, pretended none of them were satisfactory, and announced that he would take his regiment by train back to the spot where they bivouacked the night before. Instead, he had the engineer take them to New Creek, from which there was a twenty-three mile path over the mountains to Romney. The Eleventh made a forced march all night and approached Romney at dawn. The Advance Guard drove away a party of Cavalry, who rode off to warn the town that Wallace found defended by two field pieces and 1,000 to 1,200 men. Though outnumbered, Wallace outflanked the guns and attacked. There was a brisk skirmish as Wallace led the First Company under enemy fire across a covered bridge over the south branch of the Potomac. While the Zouaves scaled a rocky gorge, the Confederates withdrew, abandoning their arms, equipment, and supplies. Taking the plun-

The Eleventh Indiana in Indianapolis, May 1861, swearing to "Remember Buena Vista".

der with them, Wallace's Zouaves marched back to the train and were in Cumberland before midnight.

The episode was exceedingly well timed to give a much-needed boost to Union morale. General Patterson asked for a full report, Winfield Scott sent a personal letter of congratulations, and Indiana Congressman Schuyler Colfax wrote to Wallace that

Colonel Wallace and staff, about the time of the "splendid dash on Romney" (1861).

President Lincoln had personally commended the "splendid dash on Romney." Wallace's fame spread when newspapers played up the affair and an artist from Frank Leslie's *Illustrated Weekly* visited the camp and did a full page engraving of Wallace and his staff on horseback. Winslow Homer, the greatest artist of the war, was a young illustrator for *Harper's Weekly* in June 1861, when he visited Wallace's encampment and did five pictures of Zouave activities for the magazine plus a sketch of Wallace in Zouave uniform on horseback, the scarf of his Kepi billowing behind him, his saber at his side, and his right arm thrust forward directing the action. Colonel Lew Wallace was becoming a household name.

Major General Lew Wallace, at 34, was the youngest Union officer to hold that rank.

Private Al Fobes distinguished himself in the Romney skirmish and wrote to his parents that he had been promoted to First Sergeant and Fifer in the Regiment. He also related that his sidearm, the trusted M1849 Colt Pocket Pistol, had figured importantly in the hand-to-hand fighting that occurred in Romney.

During the war while hunting game, Al Fobes accidentally blew off the little finger of his right hand. Fortunately, this was his only injury during the war, but he did suffer much privation that was to cause chronic rheumatism, heart and nervous system disease, lumbago, kidney disease and general debilitation, as documented in his claim for pension in May of 1891.

Later in the war, Fobes was listed as officer bookkeeper, Western Division, Army of the Cumberland, Headquarters District – Nashville, Tennessee. Al Fobes was mustered out June 26, 1865, at Nashville and returned to Franklin Township, Wayne County, Indiana, in search of a job. He found work as a railroad accountant for the Illinois Central Railroad, where he remained employed from 1866 to 1889. Unfortunately, Fobes's parents had lost their farm and home in the financial panics of the 1870s, so the Fobes family was broke.

In 1890, he retired to Indianapolis, Indiana, as an invalid under Pension Certificate #732885, which entitled Al Fobes to only $8 per month.

Being somewhat destitute, it appears Al Fobes ran into problems with the law per the following newspaper story:

> ***Albert D. Fobes Surrenders, Telling A Pitiful Story***
>
> Albert Fobes, 60 years old, a paroled convict from the Indiana State Penitentiary, surrendered himself to Chief of Detectives, Mr. Desmond, last night and was placed in the holdover cell at the Four Courts, pending communication with the authorities at the Indiana State Prison.
>
> The old man, bent with age, told a pathetic story to Chief Desmond when he requested to be locked up, appearing to have had enough punishment for his misdeed. Formerly he was a clerk in the Soldiers Home at Lafayette, Indiana. While thus employed he forged a check for $42.95. His theft was discovered and he was arrested, tried, and convicted. He began his prison term April 11, 1900 at Michigan City, Indiana, where the state prison is located. After serving two years and three months, he was paroled to work in a manufacturing establishment in Michigan City.
>
> The old man was informed he would have a clerk's job, but when he arrived he says he was put to work at hard physical labor. He could not stand it and left, going to St. Louis. Finally he ran out of his meager savings and decided to surrender. He is willing to go back and serve time for violating parole rather than be free but homeless and penniless. This is a sad story to go from a valiant veteran to a penniless prisoner.

From 1894 to 1905 off and on, Al Fobes lived at the Central Branch, Old Soldiers Home, Indianapolis, Indiana. He died peacefully in his bed June 26, 1905.

Chapter Seven

The Coffey Brothers of Virginia and Their Confederate Colt M-1860 Army

I had been searching quite awhile for an identified or inscribed 1860 Colt Army with the potential of an interesting story, because I did not have that model in my collection. Finally, one turned up at the National Gun Day Show in Louisville, Kentucky, but another collector had purchased it just as I arrived at the show. Sometimes it is a blessing to miss out on something you wanted!

A few days later, my friend, Charles Layson, owner of Antique and Modern Firearms in Lexington, Kentucky, called me in Columbus saying he had just taken in a revolver on consignment that he thought I should buy (since he knows my interests). Charles said the owner, who lived in Knoxville, Tennessee, had inherited the gun from his grandmother, Mary Russell Coffey, when he was ten years old and, now in his seventies, he thought he should sell it. The revolver originally belonged to his great-great-grandfather on his mother's side of the family, Charles Edward

Charles Edward Coffey's Colt Model 1860 Army with Holster, Cap Box, and Bullet Mold.

Traffic in rodents.

The Barrel shirt.

Coffey. He was a Confederate soldier from Alto, Virginia, and had fought with General Stonewall Jackson in the Shenandoah Valley campaign from May-June 1862. He also fought alongside General Robert E. Lee against General Grant at the Battle of the Wilderness in May 1864 outside Fredericksburg, Virginia.

The seller had signed a notarized affidavit as to the truth of this provenance on July 12, 2012. So far, I was thrilled with the information, but what was the description of the gun? Charles stated it was an 1860 Colt Army Civilian Model, Serial #149149, manufactured in 1862 and in fine condition. I about fell off my chair, because this gun was just what I had been looking for. So, I bought it over the phone.

I had received permission to call the owner to help me out with the history of this Colt. When I called, Ken told me the gun had hung in his bedroom until he was 24 years old. Then he moved to his own place, where he had stored the gun in his closet through the years until he decided to sell it. Next, Ken gave me a huge surprise — he said he had found the original holster, cap box, and a bullet mold that he would send me at no charge along with a lot of family history. This was a wonderful gift and far more than I expected. So, I felt thankful I had missed the Union 1860 Colt Army in Louisville.

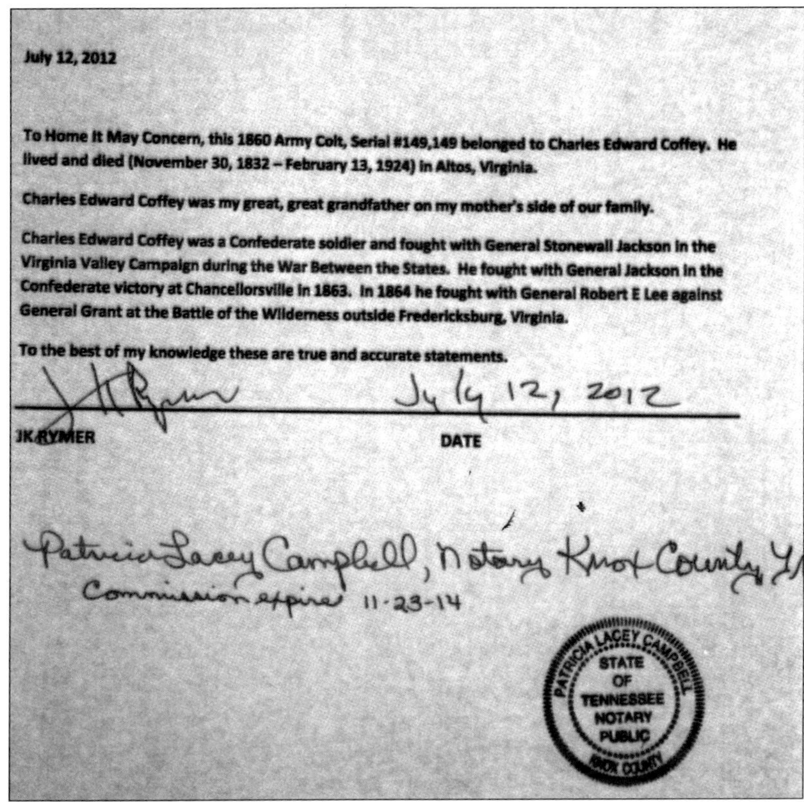

July 12, 2012

To Home It May Concern, this 1860 Army Colt, Serial #149,149 belonged to Charles Edward Coffey. He lived and died (November 30, 1832 – February 13, 1924) in Altos, Virginia.

Charles Edward Coffey was my great, great grandfather on my mother's side of our family.

Charles Edward Coffey was a Confederate soldier and fought with General Stonewall Jackson in the Virginia Valley Campaign during the War Between the States. He fought with General Jackson in the Confederate victory at Chancellorsville in 1863. In 1864 he fought with General Robert E Lee against General Grant at the Battle of the Wilderness outside Fredericksburg, Virginia.

To the best of my knowledge these are true and accurate statements.

JK RYMER — July 12, 2012 — DATE

Patricia Lacey Campbell, Notary, Knox County, T
Commission expires 11-23-14

PATRICIA LACEY CAMPBELL STATE OF TENNESSEE NOTARY PUBLIC KNOX COUNTY

Letter of authenticity from J.K. Rymer certifying that 1860 Army Colt Serial #149149 belonged to Charles Edward Coffey, his great-great-grandfather.

Ken's family history package was extensive, containing a photograph of Charles Edward Coffey, circa 1910 when Charles was 78 years old, and a photo of Edwin H. Coffey, Ken's great-grandfather. As it turns out, there were many heroes in this family.

John and Betsy Coffey from Virginia were the parents of four sons: Charles Edward, John J., William Henry and Daniel R. Coffey. All four sons were volunteer Confederate soldiers. All became prisoners of war in various Union prisons during the last year of hostilities. The happy ending is they all walked home without extensive injuries, although William developed some paralysis on his left side. Can you picture walking home from New York or Delaware down to Virginia? Put yourself in place of the parents, John and Betsy. Then, try to imagine the fear, concern, and turmoil that would grip you as all four of your sons, aged fifteen to twenty-eight, marched off to an uncertain and bloody war. One consolation is they did have two daughters, Mary and Sarah, who remained home to help run the farm.

Charles Edward Coffey was born in Amherst County, Virginia, on November 30, 1832, and died at 91 years old on February 13, 1924. He is buried in Alto in the cemetery at the forks of County Route 633 and 634, as are two of his brothers. His gravestone is large befitting a respected, kind, and prosperous man. His wife, Sarah Jane Ogden, is buried next to him. She was born October 22, 1833, and died twenty years before her husband at the age of seventy on January 12, 1904. They had seven living children: Hiter, Editha, Eleana, Mary Jane, Charles, Pitiward, and Edwin H. Coffey. Edwin was the father of Mary Russell Coffey, who left the 1860 Colt Army to her grandson, Ken Rymer.

Charles and his three brothers were Virginia Militiamen prior to the war. Charles and John enlisted together on July 1, 1861, both as Privates in the 13th Regiment Virginia Infantry, 2nd Co. E under the command of Colonel Ambrose P. Hill.

Charles and John fought the following battles together, including: Newcreek, West Virginia, June 19, 1861; Lewinsville, Virginia, September 11, 1861; Shenandoah Valley Campaign, May/June 1862; Cross Keys, June 8, 1862; and Fort Republic, June 9, 1862.

On or about June 30, 1862, Charles and John deserted near Richmond and returned home to care for their sick families and run-down farms. Desertion was common in the Civil War and did not carry the stigma that it would in today's modern military. After the two brothers took care of their affairs at home, they re-enlisted in the 50th Virginia Infantry Regiment in September 1863. Both brothers fought in: The Bristoe Campaign, October 1863; Mine Run Siege, November/December 1863; The Wilderness, May 5th & 6th 1864; and Spotsylvania Court House, May 8th-21st, 1864.

After the Wilderness, Grant's and Meade's advance on Richmond was stalled at Spotsylvania Court House on May 8th. This two-week battle was a series of combat actions along the front. The Union attacks against the bloody angle at dawn, May 12th-13th, captured nearly a division of Lee's army and came close to cutting the Confederate Army in half. Confederate counterattacks plugged the gap; the fighting continued for twenty hours in

Charles Edward Coffey, 1910

Edwin H. Coffey, 1900, son of Charles Coffey

what may have been the most ferociously sustained combat of the Civil War. General Lee was outnumbered by Grant two-to-one (100,000 U.S. – 52,000 C.S.), and estimated casualties were about 30,000 (18,000 U.S. – 12,000 C.S.). The battle was inconclusive and on May 21st, Grant disengaged and continued his advance on Richmond. Charles and John Coffey were among the prisoners of war taken during this battle.

The two Coffey brothers were confined temporarily at Point Lookout, Maryland, and then transferred by cattle boat to another prison camp in Elmira, New York, on August 2, 1864. According to southern accounts, Elmira was the worst Union prison camp, having a death rate of twenty-five percent and only in existence a little over a year. Andersonville had a death rate of twenty-seven percent, but was in existence over twice as long. Various inmate accounts report that they were under guard by black soldiers whose conduct and treatment of prisoners was infamously cruel. The guards would fire their muskets indiscriminately into crowds of prisoners. These events were never investigated. Water for prisoners was collected in barrels, and when prisoners attempted to drink, the guards would delight in immersing the drinker's head in the barrel. This was considered fun and was a daily occurrence.

The transfer ship from Point Lookout to Elmira was a cattle transport on par with the condition of the Yankee slave ships with a cargo of human souls. The rations consisted of pork fat, bread, and water. The prisoners slept on blankets spread on the hard boards in the hold. Insufficient ventilation and no sanitation in the ship's hold created stench and disease, and many deaths resulted. Swine were accommodated better, being properly fed and on the upper deck in the fresh air, while prisoners were between decks that were intensely polluted by vomit and excrement. Upon arrival at Elmira, the Confederates were marched miles in the intense August heat.

The conduct of many of the physicians in charge of hospitals deserves special notice, and the strongest condemnation is that they could not have exhibited greater brutality. Prisoners were crowded two or three to a bunk. Nearly every patient received opium pills as treatment for the many and varied diseases. On one occasion, three prisoners on opium were visibly shaking. The surgeon-in-chief was called in. He directed the ward doctor to write four or five drops of Fowler's Solution of Arsenic. He mistakenly wrote *forty-five*, killing the patients in a very short time. No investigation ensued. Not one in five of those in the hospital left alive.

Union prison camp in Elmira, New York

During the cold winter, some of the prisoners had to be housed in tents, and many suffered frostbite. Some inmates built little mud chimneys so their tents could have a bit of heat, but during the night heartless guards would kick over the chimneys, leaving some soldiers to freeze to death. There was no coffee, no tea, no meat, and no vegetables but a few beans to make tasteless watery pork fat soup…no wonder prisoners relished eating dogs, cats, and rats. A regular business of rat dealing grew up. The inmates had no opportunity to cleanse themselves of vermin. In several instances, contagious diseases were introduced into crowded prisons. Hundreds died as a result of smallpox introduced by patients from Blackwell's Island, New York. No comfortable buildings were provided for the diseased, even when the temperatures dropped to -20° below zero.

The outrageous manner in which men were vaccinated surpassed savagery. Medical assistants would take hold of a thick piece of flesh, dip a lancet into the vaccine, then thrust it entirely through the pinched-up flesh. Fearful results occurred so that it became necessary to create gangrene hospitals, from which arose a dreadful stench.

An instrument of torture for minor offenses was the wood sweatbox, just the size of a man, where men were confined for hours in the heat of the summer without food or water. Many came out more dead than alive. Another instrument of torture was the dreaded "Barrel Shirt." This was a heavy barrel with one end removed, the other end having a hole cut in it a little bigger than a man's head. All offenders, twice a day for two hours, had to wear it and march around in a circle. On the back of each barrel was recorded the offense such as "Dog Eater", "Liar", "Rat Dealer", etc.

There were dozens of other prisoner abuses on both sides, but the above serves to illustrate the fearful conditions all four Coffey brothers endured. Luckily, all the brothers spent less than a year in prison. Charles and John were in Elmira from August 2, 1864 until June 22, 1865. William Henry Coffey was imprisoned at Fort Delaware from August 1, 1864 until June 19, 1865, but was delayed on his homeward mission for five months due to illness. Years later, the family told of his homecoming:

"Grandpa had given him up for dead but grandma would say, 'Hen is coming home to us, Jack.' This was grandma's true faith in God. Months later, she saw someone coming up their long lane so weak he kept stopping to rest. She started running, calling him, 'Hen.' When he got close, he said, 'Don't come near. I am unclean. Get a tub of hot water and set it out near the chimney.' After he got clean and put on fresh clothes, he was greeted. He soon gained flesh and strength. But after he got married, he developed some paralysis on his left side — but he was a wonderful provider with two sons and a loving wife."

The youngest brother, only 18 years old in 1864, was Daniel Coffey. He was captured near Appomattox in May 1864, and then was exchanged for Union soldiers on November 23, 1864, thus spending only about six months in Fort Delaware prison.

The Coffey's were certainly a family of heroes — the parents, John and Betsy, for providing the four warriors and going through the agony of uncertainty and running the farm extremely shorthanded; obviously, the four soldiers for volunteering, fighting bravely, and enduring prison, and the long walks home; and, lastly, the daughters, Mary and Sarah, who became farmhands to keep the family farm from failing.

Charles Edward Coffey and Sarah went on to have six more children after the war, for a total of ten. However, one died in infancy. Charles became a pillar of his community of Pedlar, Virginia, an Elder in his church and a successful farmer of over 1,000 acres. When Charles died in 1924, he left each of his nine children $7,000, which meant his estate was over $63,000 — a small fortune in 1924 dollars. He also released his three slaves—Billy Ogden, Joe Ogden, and Spicey Ogden—after the war, giving each of them a tract of his own land. They lived and worked this land, raised their families, and stayed in contact with Charles until his death.

Charles was indeed respectable, God-fearing, kindly, hardworking, generous, and successful — all you could ask of a man then and now!

Chapter Eight

The Bacon Revolver of John Hancock, Who Served Both the U.S. Army and the U.S. Navy During the Civil War

Could he be related to "the John Hancock" of Revolutionary War fame and a signer of the Declaration of Independence?

In April of 2002, I purchased a nearly factory-new, first model Bacon 5 Shot Revolver, .31 caliber with a 4" barrel at the renowned Ohio Civil War Show in Mansfield. This gun, Serial #102, was manufactured in Norwich, Connecticut, about 1860 by the Bacon Manufacturing Company.

John Hancock's Bacon Revolver, Mfg. 1860

Less than 1,000 were made, so this is a relatively rare model, especially in this condition. The back strap and butt strap were engraved, rather uniquely, "John Hancock, Peekskill, N.Y." as shown in the photos. At the time of purchase I knew the inscription was authentic and valid, and that this John Hancock had served in the Civil War. The reputable gun dealer knew nothing

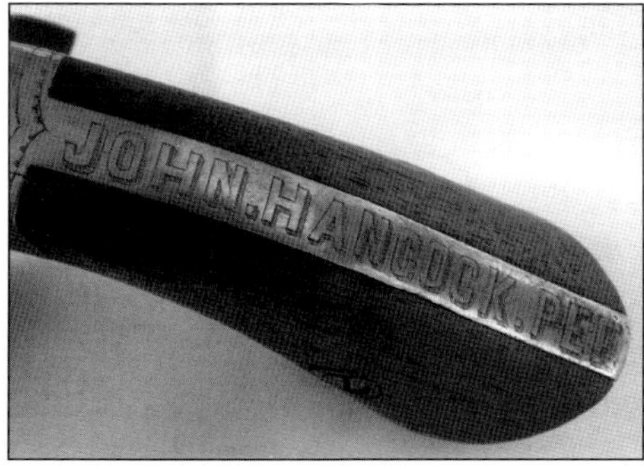

further, because he had completed no research, having just recently acquired the gun.

John Hancock

Vonnie Zullo of Fairfax, Virginia, a very skilled researcher affiliated with the Horse Soldier Civil War Shop in Gettysburg, Pennsylvania, searched the archive records in Washington, D.C. for me. I was very pleased with her findings, which verified that John Hancock had enlisted April 20, 1861, as a Private under Colonel Daniel E. Sickles (later General Sickles) with Co. I, N.Y. 70th Volunteer Infantry, which was assigned to Brigade 2, Hooker's Division, Army of the Potomac.

Several unexpected dividends came from Vonnie's Research — John Hancock had been discharged at Falmouth, Virginia, on March 30, 1863 because of a combination of typhoid fever and battle wounds. This fact provided me with his complete pension records, which contain so much additional personal information and insights into his entire life. Hancock endured a painful year of recovery from knee wounds and typhoid, and then with an apparent burst of patriotism, he re-enlisted in the U.S. Navy on August 27, 1864 aboard U.S. steam warships *Vanderbilt* and *Vermont*. Hancock held the rank of Landsman in the U.S. Navy in 1864, and was finally discharged on June 6, 1865.

The 70th New York Infantry, First Regiment of the Excelsior Brigade, saw much action as demonstrated in their itinerary as follows:

– Defense of Washington D.C., July 29, 1861.

– Skirmishes at Yorktown April 1862.

– At the Battle of Williamsburg, the 70th New York met with the heaviest loss of its service. Out of 700 soldiers engaged, the loss was 330 killed, wounded or missing.

– At Fair Oaks and in the Seven Days' Battle, the regiment was

The Battle of Fredricksburg

active and embarked August 20 for Alexandria, where it moved to support General Pope at Manassas, a fierce engagement.

In a sharp encounter at Bristow Station, the 70th lost five men, and at the Second Bull Run they lost twenty-three more. The 70th then fought in the Battle of Fredericksburg, returning afterward to its camp at Falmouth, which became its winter headquarters.

At this point Hancock was discharged, but the 70th New York was next in the field at Chancellorsville, and at Gettysburg lost 113 killed or wounded and four missing.

After enlisting in the Navy August 27, 1864, Hancock served on the Steam Warship *Vanderbilt*, which patrolled in the treacherous North Atlantic against blockade-runners op-

USS Vanderbilt

erating out of Halifax, Nova Scotia. She served on the blockade off Wilmington, North Carolina beginning in November 1864, and took part in the December 1864 and January 1865 attacks on Wilmington's Fort Fisher, that finally resulted in closing that important port to Confederate commerce. This port supplied Richmond, Virginia, capital of the Confederacy. The spring of 1865, *Vanderbilt* carried sailors to the Gulf of Mexico and towed ironclads between east coast ports. Her last war duty was as a receiving ship at the Portsmouth Navy Yard in Kitterly, Maine, during the summer of 1865.

The U.S.S. *Vermont* was built as a war ship of the line by the Boston Navy Yard. She was stationed at Port Royal, South Carolina, where she supported the risky Civil War operations of the South Atlantic Blockading Squadron. Her last wartime assignment was as receiving ship at the New York Navy Yard.

I felt very fortunate that Hancock enlisted for a second time in the U.S. Navy, because either the National Maritime Historical Society in Peekskill, New York, or the Naval Historical Foundation in Washington D.C. would very possibly have photographs of the *Vanderbilt* and the *Vermont* on which he served.

As luck would have it, I was able to obtain good photos of both ships, as well as all their history, most of which is here included. Next, I contacted the Field Library in Peekskill, New York, and spoke with Mr. Bob Boyle concerning any information they might have on John Hancock. Mr. Boyle soon sent me a real bonanza that I certainly did not expect — John Hancock's life story, including a photograph

USS Vermont

when he was sixty-seven years old, was featured in Peekskill's *Highland Democrat Newspaper* Saturday, June 27, 1903. The article was titled, "Our Portrait Gallery, Peekskill People Portrayed." Needless to say, I was thrilled to obtain all this in depth information because it is rare, indeed, to obtain a photograph of the Civil War owner of an inscribed gun.

Following is a synopsis of *The Highland Democrat*'s account of John Hancock's life, which demonstrates that not only was he an exemplary soldier and sailor, but also an outstanding citizen all of his years:

> We present this week the story of a gentleman who, while not a native of this country, has been so long with us as to be called an American and who has served his country as an American. John Hancock first saw the light of day in old England, in the province of Devonshire, and comes from a family long residents of that locality. He was born the 23rd of September 1835, and is in his 67th year. He was the son of William and Susannah Hancock. By the advice of their family physician, after losing three children, Mr. And Mrs. Hancock, when John was nearly six months old, left their native land and sailed for America. His early years were spent in New York City, the stopping place of most people who came to this country in the 1830's. His first schooling was held in the public school on Rector Street and later on Reade Street. When John was about seven years of age the family moved to Peekskill and for a time he attended the school of District No. 7, which at that time was known as the yellow school house on South street bank. Later he attended Miss Caldwell's school, where Buchanan's restaurant is located on Main Street. Then he went to learn from a Miss Carpenter, who taught school further down Main Street. After this he attended Dr. Westbrock's school on Main Street near Spring.
>
> When about fourteen years of age the young man thought he would rather work than go to school, so he hired out to clean castings in Finch's foundry while his father was away in New York City on a law suit. He worked some time there

and then engaged with David L. Seyour to learn the trade of machine molder. After serving three years with Mr. Seymour he went back to Finch's and served three years as a stove molder. In the year 1857, during a slack spell in the Peekskill foundries, he worked for a time at Tarrytown and Spuyten Duyvil. Work in Peekskill was again picking up so he came back to Finch's foundry and remained there till the war broke out.

Happening to be in New York when the Seventh regiment departed for the front, he was fired up by the sight of the soldiers and concluded to enlist. On his way home he met Daniel Clark Briggs, who informed him he was about to raise a company in Peekskill. On the 20th of April in a room hired for a recruiting office over the blacksmith shop of George Lockwood, on Diven Street, Mr. Hancock was the first man to sign his name to the muster rolls of the First Regiment, Sickles Brigade. He went to the front and served in many battles for two years and eleven months. Between Malvern Hill and the Wild Swamp he was wounded in the knee. Later in December, while fording a river, he took cold, got typhoid fever and was soon after sent home for disability. After doctoring for some time he thought the sea air would be beneficial and, still having plenty of fight left in him, he enlisted in the Navy and with nineteen comrades from Peekskill went aboard the United States Steamer Vanderbilt. They were in the two battles of Fort Fisher and came home just in time to hear of the death of President Lincoln in 1865, ten months after enlistment.

After coming home Mr. Hancock worked for a short time and then became manager of the Peekskill Enameled Iron Works. After two years there the company failed and then Mr. Hancock, in the winter of 1868-69, was appointed principal doorkeeper of the assembly at Albany. In the fall of that year he was appointed to a position in the New York post office, in charge of the western department. He stayed there for nearly seven years. For two years after this Mr. Hancock was a warden in Sing Sing prison. He came back to his trade again and made and mounted patterns for the

Union Stove Works for a time. He, in company with some other Peekskill men, went to Quincy, ILL, and worked for a month in the foundry of Bonnett & Duffy. Then he became manager for the Quincy Stove Works. While there he was ill for some time and his life was despaired of, but he recovered enough to come back to Peekskill after having been away not quite a year. About a year after he came home he went to work for Ely & Ramsey making patterns, and later for Geo. D. Sanford.

In 1888 Mr. Hancock secured a position as manager of the new capital at Albany. Here he remained for about 12 years. In 1895 he opened a grocery store on South street in which he continued for about one year. He then acted as court officer at White Plains for some time. In semi-retirement, Mr. Hancock took the position of facilities manager of the Depew Opera House till the building burned in 1901. He was then appointed facilities manager of the New Franklin Street School.

In 1854, Mr. Hancock married Miss Abbey Denike. They had two children, one of which died in infancy. His wife died May 31, 1858.

In 1861 he married Miss Annie Brotherton in New York. They had one daughter. Mrs. Hancock died January 5, 1874.

In 1888 he married Miss Katie Hirst in New York, and they now reside at 415 South Street.

Mr. Hancock is well up in the charitable fraternities. In 1856 he became a member of Cortland Lodge No. 6 I.O.O.F. (Independent Order of Odd Fellows) and is the second oldest member of that organization now living. He served in all the chairs and became a Past Grand Master. He also passed through all the chairs in Mt. Ararat Encampment, I.O.O.F. For many years he was a member of Iron Molders Union No. 6 and is now an honorary member of that organization. Mr. Hancock has also been a prominent Mason for many years. He was raised in Cortland=s Lodge No. 84, Free Masons, August 14, 1867. He is also a member of Mohegan Chapter No. 231, R.A.M. (Royal Arch Masons), and Peekskill Council R.&S.M. (Royal & Select Masters) No. 55, and holds the office of sentinel in each. He is also a member of Manhattan Commandery, No. 31, K.T. (Knights Templar), and is a 32nd degree Mason.

In politics Mr. Hancock has always been a straight Republican and has done good work for that party for many years.

Mr. Hancock is of a reserved nature, but with those who are intimate with him he is a genial, whole-souled man who makes friends wherever he goes and keeps them after he makes them. May he add many years to the already large number he has lived in the enjoyment of good health and prosperity. John Hancock lived most of his life in the Peekskill, Westchester County, New York area. In March 1897, John Hancock lived at 718 John St., Peekskill, NY and was 62 years of age. In June 1903, his residence was 415 South Street, Peekskill, NY and by September 1905 he and his wife, Selina, lived at the corner of Frederick and Franklin Streets in Peekskill. He was 70 years old and unable to work any longer. His pension was $12 a month. At age 75 in 1910 his pension was increased to $20 per month, in 1912 to $30 per month and in 1918 to $40 per month. John Hancock died November 12, 1919 at 84 years of age and was buried at Hillside Cemetery, Peekskill, NY. His widow, Selina died September 8, 1935 at 87. Her residence was 508 Van Cortland, Park Avenue St., Ward 8, Yonkers, New York.

How unfortunate would we feel, in 2020, to lose a wife and child after only four years of marriage, and then lose your second wife and be left alone with two children to raise? Our life today must be so much easier on our spirit, or is it?

Regarding the potential of the Civil War John Hancock being related to the Revolutionary War John Hancock, this issue is currently being thoroughly investigated, because there are many preliminary indications that they are indeed relatives:

* Both Hancocks immigrated to America from the Devonshire, England area.

* Comparing portraits of both men, there is a striking resemblance.

* Both Hancocks were exceptionally patriotic and both were outstanding citizens.

* Boston, Massachusetts is only one hundred ninety miles from Peekskill, New York, so both Hancocks lived in the same general area.

In summary, I hope to unearth concrete proof of their family relationship, but this matter will take much time and must await additional research.

Chapter Nine

Factory Engraved Colt M 1851 Navy Presented To Colonel John Warner

Guns can often tell great stories, especially when they are inscribed, which offers the opportunity to do historical research.

In 1993, I purchased this Colt Navy at the Baltimore Gun Show (Maryland). Before completing the purchase, I showed the engraved inscription to several gun experts, who confirmed it was legitimate. The only paperwork accompanying the pistol was copies of several reports that Colonel Warner had filed with his superior officers.

I really enjoy doing historical research, so I immediately sent for Col. Warner's archive records, which indicated he enlisted August 11, 1862, at Peoria, Illinois, elected Col. August 26, 1862, and mustered out March 13, 1864. The 108th Illinois was attached to the 2nd Brigade, 1st Division, Army of Kentucky and saw action with Sherman's Yazoo Expedition, Chickasaw Bayou, the Capture of Fort Hindman, Arkansas Post Battle of Port Gibson, Mississippi, Champions Hill, and the Siege of Vicksburg.

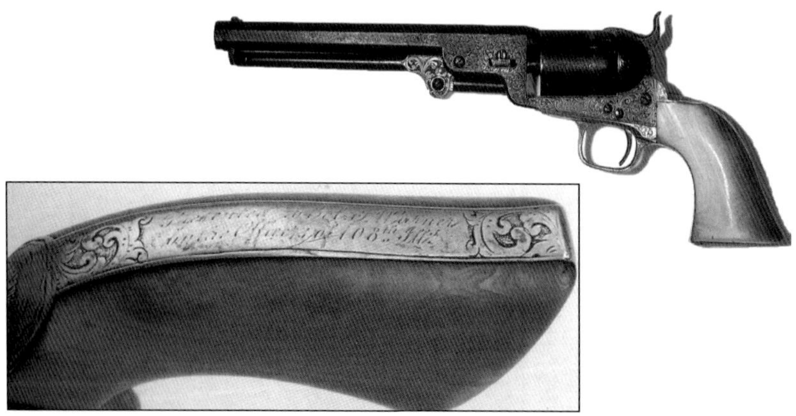

Col. John Warner's factory engraved Colt M 1851 Navy. The inscription reads: Presented to Col. John Warner by the officers of the 108th Illinois.

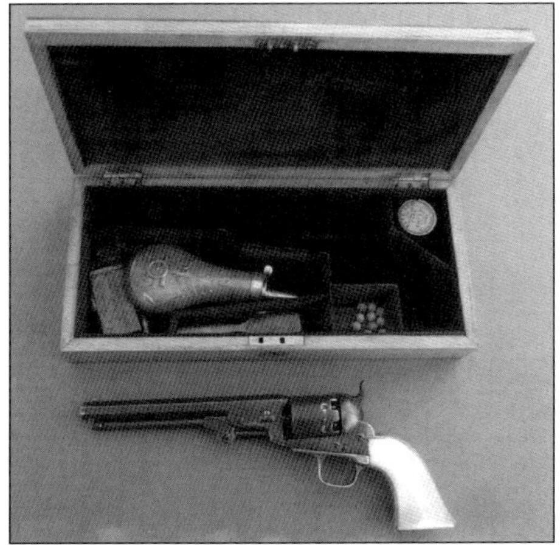

The Colt M 1851 Navy is cased with accessories.

Col. John Warner

Excited by his military records, I contacted the Bradley University Library, the Peoria Public Library, and the Peoria City Hall. To my delight, I discovered that John Warner had been Mayor of Peoria and that his picture was hanging in the Council Chambers. I soon arranged for a personal visit to Peoria, where I was able to unearth a great volume of data on John Warner and his family.

For thirteen years, intermittently between 1874 and 1898 (inclusive), John Baptist Warner, a Democrat, served as Mayor of Peoria. His administrations were marked with considerable achievement.

John Warner was born in Perry County, Marion, Ohio, in 1826, the son of John B. Warner, a native of Maryland, and Esther (Gordon) Warner, a native of Pennsylvania. The Warner's were descended from an English family, which took root in America as early as 1657. The great-grandfather of George Washington came to America at the same time. The two families were acquainted and connected by marriage. Col. Warner's grandmother, Betty, was a sister of George Washington and lived just across the Potomac River from Washington's plantation.

John B. Warner, the father of our subject, Col. John B. Warner, was a contractor who engaged in both railroad and canal construction. He was also experienced in hotel operation. Mr. and Mrs. Warner Sr. were the parents of six sons and two daughters.

In 1846 the Warner family moved to Peoria, where the father, in 1849, became owner and proprietor of the historic Clinton Hotel, which stood on the corner of South Adams and Fulton Streets, where the Block & Kuhl Company building now stands. Mrs. Warner was an excellent cook. Mr. Warner was a Justice of the Peace as well as hotel proprietor. Young John B. Warner became a stagecoach driver that traversed the region, arriving, departing, and changing horses at his father's hotel.

At this time, young Warner met Abraham Lincoln, who often stayed at his father's Clinton Hotel. Lincoln and Warner became friends and young Warner went to hear the famous debates between his friend Lincoln and Douglas.

In 1854, Col. Warner married Miss Elizabeth Simms of Peoria, daughter of Mr. and Mrs. Alonzo Simms. In 1858, he entered the mercantile business on South Water Street, featuring men's clothing. He also managed several riverboats and engaged in the ice business. In February 1860, he received a contract to deliver three thousand tons of ice to New Orleans by steamboat.

Following the outbreak of the Civil War, he entered the Northern Army, serving as Colonel of the One Hundred and Eighth Illinois Infantry. The regiment under his command served with distinction in several important engagements, among them Chickasaw and Vicksburg. Colonel Warner often told Civil War stories, including how he had killed two Confederates with his Navy Colt in hand-to-hand combat.

When the war was over, Col. Warner was for two years with the U.S. Revenue Service. Then, for seven years he engaged in the wholesale liquor business. He was elected to his first term as Mayor of Peoria in 1874.

During his six terms in office, a number of major improvements were made. In 1879, the Peoria workhouse was established. In 1880, Mayor Warner nominated the first Board of Directors to serve the Peoria Public Library, under the Illinois Free Public Library Act. In 1897, during his last term in office, Peoria's third City Hall Building was erected at Fulton and South Madison Streets. While he was Mayor, a reorganization of the fire department was effected, and uniforms were provided to the police department. In 1883, Mr. Warner served as postmaster, resigning at the end of one year. Mayor Warner's secretary recalled that he did a great deal of target practice, and always kept his Navy Colt in his office for personal protection. He also often showed his Presentation Colt to visitors because of his enormous pride in serving in America's Civil War.

Colonel and Mrs. Warner became the parents of eight children: John A., Dollie, Harry, Etta, Agnes, Molly, Daisy, and Cora.

John, the elder son, married Miss Pauline Meidroth of Peoria. He was for some years a member of the Peoria Fire Department and once served as assistant fire marshal. Dollie married Frank Buell of Peoria. Harry, who remained single, moved to Colorado. Etta married William D. Meisser of Peoria, who was long associated with the linen department of Schipper & Block and the succeeding Block & Kuhl Company. In 1887, he served as city controller. Agnes married Frank Dever of Peoria. Molly married H.J. Humphry of Chillicothe. Daisy married Joseph Hall of Peoria. Cora married Charles Geihausen of Peoria.

The early Col. John Warner residence was on the upper side of the one hundred block of South Monroe Street. Before 1870, Col. Warner erected a pleasing two-story frame residence at 105 Third Street, where the family resided for twelve years.

Through the courtesy of a granddaughter, Mrs. M.A. Hoag, I was able to see a photograph of the Third Street residence. At the old hitching post in front of the house is Molly O, the Colonel's highly prized horse. Molly O had a good racing record. About 1882, the Warner family moved to 203 Park Place.

Of Mr. and Mrs. Warner's grandchildren, only Mrs. Peoria Meisser Hoag lived in Peoria. She is the daughter of Mrs. Etta Warner Meisser. Two great-nieces of Mrs. Warner, Mary Irene and Eva Wonder, also made Peoria their home.

When the first Federal Housing Project in Peoria was established and the first unit completed in 1941, near the South Side, it was named "Colonel John Warner Homes" in compliment to the former Peoria Mayor.

Col. Warner's death occurred on December 28, 1920 at the age of 94. Mrs. Elizabeth Simms Warner, his wife, died July 11, 1916. Peoria honored Col. Warner many times including his death. His body lay in state at City Hall from 9:30 A.M. to 1:30 P.M., on Friday, December 31, 1920.

Chapter Ten

Old Friends Meet Again – A Tale of Two Smith & Wesson Old Army #2s

In 1997, I acquired Smith & Wesson Old Army #2, SN 8621 from Dave Taylor of Civil War Antiques in Sylvania, Ohio. At the time I knew the inscription: "Presented to J.F.E. Chamberlain by his friends at the U.S. Armory" was unquestionably correct. I also knew that Chamberlain had served in the Civil War and later was the principal sub-inspector on the first 3,000 Smith & Wesson Schofield revolvers completed June 29, 1875. His initials J.F.E.C. appear in script on the left grip of many first model Schofields.

Chamberlain's #2 Old Army has a silver-plated frame and blued barrel and cylinder. According to Roy Jenks, Smith & Wesson's historian, this gun was shipped in September 1862 and

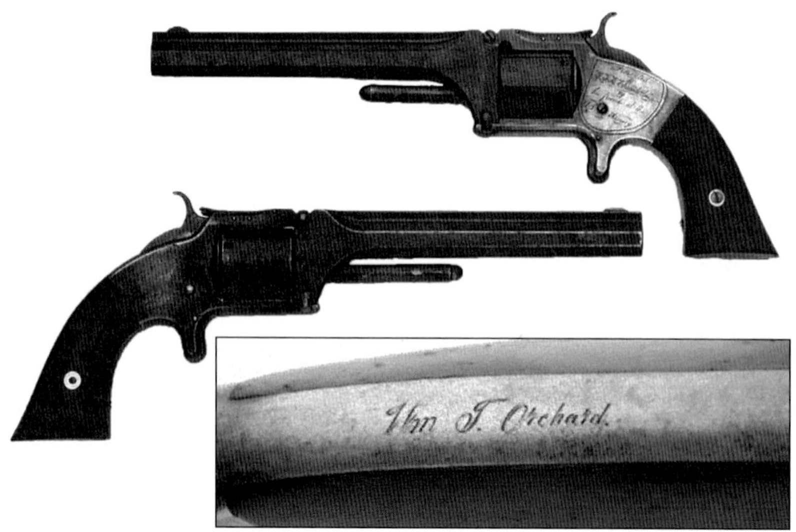

The Chamberlain Civil War revolver shown above the Orchard Civil War revolver. Inset is the signature of William Orchard as seen on his pension application. Below right is the "Wm. T. Orchard" inscription on the lower revolver.

delivered to the Wm. Patton Co., 102 Main Street, Springfield, Massachusetts. Mr. Patton was a jeweler and employed engravers who probably did the inscription, which is on the left side of the frame.

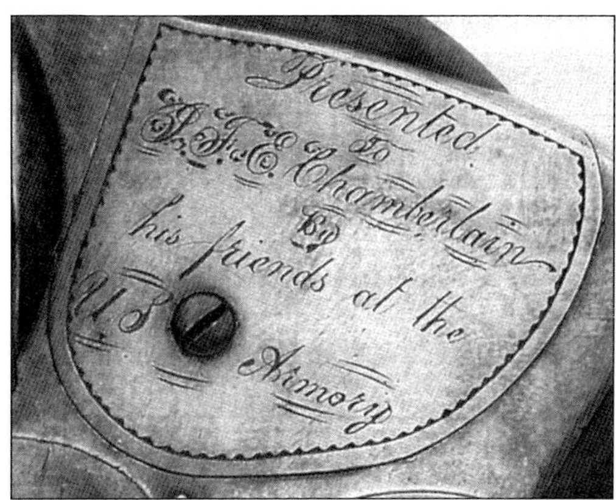
The handle of the Chamberlain Civil War revolver reads, "Presented to J.F.E. Chamberlain by his friends at the US Armory."

Dave Taylor had said Chamberlain served in two Civil War units — the 46th Mass. Infantry and the 3rd Mass. Heavy Artillery, so I decided to send for his archive records. From these I learned John Fitz Edward Chamberlain was born October 15, 1838, in Cavendish, Vermont. He stood 5'11" tall with fair complexion, hazel eyes and brown hair. Chamberlain was an armorer at the Springfield Armory before the war.

Chamberlain entered military service on September 25, 1862, as a Private, Captain Spooner's Company, 46th Mass Infantry. This company mustered in July 29, 1863. He re-enlisted on January 21, 1864, as a Second Lieutenant, 3rd Mass. Heavy Artillery. Chamberlain was promoted to First Lieutenant on February 23, 1864. While on duty at Richmond, Virginia, he contracted typhoid fever and malaria, from which he almost died. At war's end, John Chamberlain mustered out with his unit on September 26, 1865, and soon returned to his work at the U.S. Armory.

After the war, he (Chamberlain) married Martha Cutler, who died on December 26, 1877, only two years after giving birth to their daughter, Mattie. He then married Harriet Norton on November 26, 1879, with whom he had his second daughter, Lurena, born November 1, 1884. J.F.E. Chamberlain died June 4, 1923, at 85 years of age.

Then came a shock! As I researched the details of Chamberlain's declaration for additional invalid pension as a result of his malaria and typhoid fever complications, I discovered that William T. Orchard had witnessed Chamberlain's signature on June 10, 1904. The reason for my great surprise is that in my collection is another Smith & Wesson Old Army #2 inscribed: "William T. Orchard" on the backstrap. Incredibly, J.F.E. Chamberlain and William T. Orchard knew each other rather well in Springfield, Massachusetts, and I, by unbelievable chance, now owned both of the revolvers they carried in the Civil War.

I was excited, so I immediately got out my records on the William T. Orchard revolver. I had purchased this gun from Charles G. Worman, author of the two-volume work, *Firearms of the American West 1803-1865*, on November 7, 1992. Chuck Worman told me Orchard had been a Smith & Wesson employee prior to and after his service in the Civil War. It is almost certain Orchard had this gun made for his service in the war, because of the inscription and also a customized feature — the ejector rod is about 9/16" longer than normal.

Orchard enlisted in the Union Army in July 1864, at which time he listed his occupation as "Pistol Maker." He was assigned to the 30th Co. Mass. Heavy Artillery. This unit was assigned to duty in the defense of Washington, D.C., during the time Lincoln was assassinated, until the unit was mustered out in July 1865.

In reviewing my old research, I was looking for the connection to discover how Chamberlain could have known Orchard well enough to ask him to be a witness on his pension application.

The answer came in a research letter I had requested in 1996 from the Connecticut Valley Historical Museum in Springfield, Massachusetts, which reads in part as follows:

> "Thank you for your recent inquiry regarding William T. Orchard. William was the son of Charles and Lois Orchard. William first shows up in Springfield's 1861-62 Directory as an adult member of Charles' household. He was listed as an employee of Smith & Wesson. William married Laura M. Rice on February 27, 1863, about year and a half before he enlisted in the Army in July of 1864. William and Laura's daughter, Nettie Chapin Orchard, was born November 14, 1866, a little over a year after William's return from the Civil War. Nettie was their only child.

> "Orchard worked for Smith & Wesson until 1882, at which time he went to work for S.G. Otis & Company, a printer, until 1884, at that time William became a machinist for the U.S. Armory, and remained employed there at least until 1910."

The mystery was solved, since both Chamberlain and Orchard worked together at the Springfield Armory from 1884-1910. Orchard witnessed Chamberlain's signature June 10, 1904.

Inscribed Smith & Wesson's of Civil War vintage are quite rare. The fact that their respective owners were employed by such historically significant places like the Springfield Armory and the Smith & Wesson Company is even more rare, and now that we have proof that the two men knew and apparently liked each other, is truly an incredible coincidence. This is a great example of the fun and luck that comes from researching historical arms.

"Together Again!"

Chapter Eleven

John H. Masonheimer A Civil War Hero Who Got A Second Chance

In 2002, I bought a Smith & Wesson Model 1, second issue inscribed on the left side plate "J.H. Masonheimer from the battlefield of Stones River December 31, 1862". The dealer who sold it to me, Mr. Neil Gutterman, is a man of outstanding reputation who assured me the inscription was a valid period engraving; however, he knew nothing further. I thought the inscription was very unusual, in that it would be rare for an individual to inscribe a gun to himself that was found on the battlefield. Instead, someone else must have inscribed the gun to J.H. Masonheimer as a battlefield memento. I would soon learn the very interesting circumstances that led to this rather rare inscription.

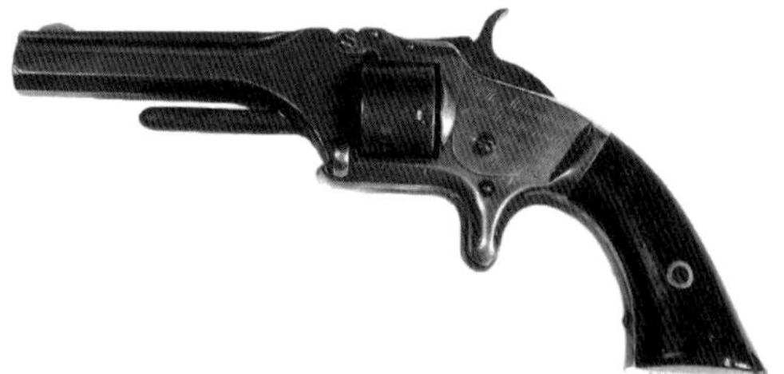

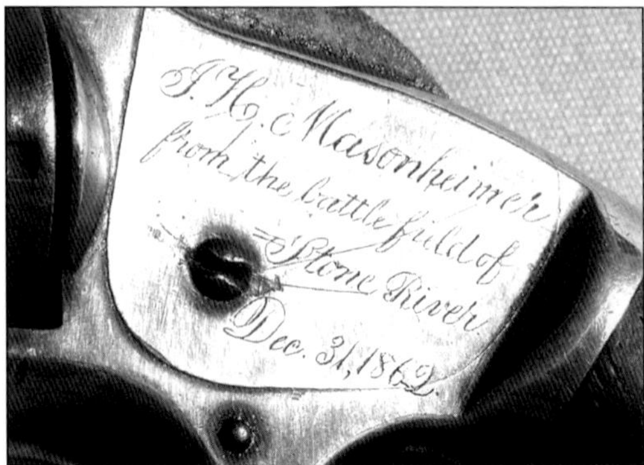

The inscription reads "J. H. Masonheimer from the battlefield of Stone River December 31, 1862".

I enjoy my hobby and find it interesting and fulfilling to research these historic pieces and the lives attached to them — it's

Main Street, Shelbyville, Kentucky.

a rare inscribed or identified gun that does not have something worthwhile, from a human point of view, to research. Doing the research is methodical, doesn't appeal to many people, and often is not particularly profitable. But to me, it brings the guns alive and adds exponentially to the hobby or business of collecting. The procedure of investigating a piece's history is rather uniform, but also embraces a bit of creative thought:

1. Check the national archives and state archives for the potential war and pension records of the individual.
2. Assuming records are available, check the subject's city and county courthouse records, historical society records, and genealogy of the individual.
3. Call the area where the subject lived and search for living descendants who may have additional information and stories about their ancestors.
4. Contact local libraries for information on the person, and sometimes local clubs may have additional information;

for example, the "Descendants of Union Soldiers Groups" or some other little known local club.
5. Search Internet genealogy and Civil War sites.
6. Visit the area personally and ask around for historical information. As the search continues, it's amazing how one thing leads to another and suddenly a fairly thorough picture of the subject develops. Sometimes I end up knowing more about a stranger than I know about my own great-grandfather.

In the case of Pvt. John H. Masonheimer, I found the following basic archival information:

JOHN H. MASONHEIMER Private, Company A
15th Kentucky Infantry
All recruits for this regiment came from Louisville and the surrounding area.
BORN: 1813 near Baltimore, Maryland
HEIGHT: 5' 9"
COMPLEXION: Light
EYES: Hazel
HAIR: Dark
WEIGHT: 145 lbs.
MUSTER IN: Enlisted October 1, 1861 at Camp Sherman near Shelbyville, Kentucky and mustered in December 14, 1861 at Camp Pope, Kentucky for a three-year enlistment.
MUSTER OUT: Honorable Discharge, January 24, 1863 at Murfreesboro (Stones River), Tennessee
RESIDENCES: Shelbyville, Shelby County, Kentucky until February 1872
Paducah, Kentucky until August 1873
Louisville, Kentucky until September 1876
Shelbyville, Kentucky until September 1877
Louisville Kentucky until August 1881
Chattanooga, Tennessee until December 1882
639 7th Street, Louisville, Kentucky in December 1883
621 7th Street, Louisville, Kentucky until October 1895
WOUNDED OR SICK: Disability Cert. #289223, Applica-

tion for Pension #497025. Received $8 per month from October 5, 1883 to 1891. He received $12 per month from 1891 until his death in 1895. His disability was due to inflammation of the lungs, chronic rheumatism and chronic diarrhea.

DIED: October 9, 1895 at eighty-two years of age

TIMELINE OF HIS LIFE: Born 1813, Baltimore, Maryland
- 1830, Married in Maryland at seventeen years of age
- 1834, At twenty-one years of age still in Maryland
- 1845, John H. Masonheimer's wife died leaving John with a fourteen year-old son, Walter
- 1861-1863, Served in Civil War
- 1875, Walter Masonheimer, his son, dies in California
- 1895, October 11, John H. Masonheimer is buried in Cave Hill National Cemetery, Louisville, Kentucky

Through contacting Deborah Magan at the Shelby County Public Library, I found that a history of the 15th Kentucky Volunteer Infantry was available, and she copied it and sent it to me. Further, she referred me to Ms. Betty Matthews of the Shelby County Historic Society, who immediately knew John H. Masonheimer as a Shelbyville citizen of importance. Betty Matthews sent me a book called *Some Old Time History of Shelbyville and Shelby County*, which had some great accounts of J.H. Masonheimer. But, I'm getting ahead of myself and must return to the time of his enlistment — October 1, 1861.

At the time of his enlistment, John was a forty-eight year-old widower with one thirty year-old son, Walter Masonheimer. He was a well-established merchant tailor, having his shop on Main Street in Shelbyville. However, he had a burning patriotic desire to serve his country by enlisting in the Union Army. He was a strong believer in the value of a united free country, and could not tolerate the thought of continued and growing slavery nor a Confederate States of America. Why else would a forty-eight year-old man enlist in an army of much younger men averaging only twenty something years of age.

On December 14, 1861 at Camp Pope, Kentucky, John H. Masonheimer was mustered into the 15th Kentucky Volunteer Infantry, Co. A – Marion C., Taylor, Captain, James A.T. McGrath and Frank Winlock, Lieutenants.

The men of this regiment all came from Louisville and its vicinity. The regiment was organized at its camp in the fairgrounds near the city, and was immediately ordered to duty on the Nashville railroad. For a short time it encamped at New Haven, and thence went to Bacon Creek, where it was formally mustered into the service.

From Bowling Green it traveled to Nashville, Franklin, and Mitchellville, Tennessee, and Huntsville, Alabama. In August 1862, it entered upon the memorable march of Buell's Army from Tennessee to Louisville. It was engaged in the severest part of the battle of Perryville. Here the battle raged from about one o'clock p.m. until dark; the rail fence near the Union troops was almost totally demolished by the enemy's artillery, and a nearby barn was set on fire by shells; however, despite this bombardment, the fifteenth held its ground.

The color guard, consisting of nine sergeants, was cut to pieces. The fight was so fierce and continuous that the colors were completely riddled with shot holes, and the flag staff was cut in two.

As the staff was severed and the colors fell, Captain James B. Forman of Company C grasped them, and as the staff had been cut off so short that they could not be made visible, he mounted the remains on the rail fence, waving them, cheering the men to continued resistance. The Fifteenth Kentucky retained its origi-

Perryville, Up the hill came the Rebels.

Confederate attack on Harris's brigade. From Harpers Weekly, November 1862. Photograph by Tom Beggs

nal line until the close of battle, and in the morning the enemy was gone.

Our first full-fledged battle was a "Baptism of Blood" and it served to cement the regiment more closely in love one to another and to the cause for which it fought. Sixty-three men were killed including three officers, and two hundred were wounded. In late 1862, Pvt. Masonheimer fell ill with inflammation of the lungs and rheumatism, and was ordered to the 15th Kentucky Regimental Hospital. Masonheimer confided to his good friend, Lieutenant James McGrath, that he was terribly distressed to be hospitalized when important battles were likely to come soon. Lt. James McGrath was the nephew of Thomas C. McGrath, who was a merchant in Shelbyville located near J.H. Masonheimer's Tailoring Shop on Main Street.

Prior to the Civil War, Thomas McGrath and John Masonheimer were well acquainted and quite friendly, since they were fellow merchants in Shelbyville and belonged to the Masons and other businessmen's groups. Naturally, his nephew, Lt. James McGrath, was concerned for Mashoneimer's health and promised to do his best to get Masonheimer back into the field, since he was quite aware of his intense patriotism and eagerness to serve the Union cause.

After the Battle of Perryville, the 15th Kentucky continued to Nashville, where it was assigned to Beatty's Brigade, Rousseau's

Division, Army of the Cumberland, commanded by General George H. Thomas. It was engaged in the great Battle of Murfreesboro, or Stones River, where Colonel Forman and eighty others were either killed or wounded. There, in three days of fighting between the Federal Army of the Cumberland and the Confederate Army of Tennessee, 20,000 men had fallen with no advantage to either side.

In the Words of William P. McDowell:

> "To sum up the history of the Fifteenth Kentucky is a task both pleasant and painful.
>
> "Pleasant because it can be said that in our whole service we were almost always in the front lines of the army, and always received the commendations of our commanders and the love and esteem of our companions.
>
> "Painful, from the fact that of eight-hundred and eighty-eight men and officers mustered into the United States service in 1861, over four hundred were killed and wounded on the battle fields of our country."

Unfortunately, Masonheimer's condition grew worse, and he was forced to remain in the regimental hospital through December 31, 1862, thus totally missing the great Battle of Murfreesboro (Stones River, Tennessee).

Battle of Stones River

Lt. James McGrath, being a man of great sensitivity, realized the anguish of his friend Private Masonheimer and thought to bring an uplifting present to his bedside in the hospital. On the battlefield, McGrath had found a Smith & Wesson Model 1, 2nd issue lost by a soldier, and thinking to bring his seriously ill friend a memento of their victorious battle, had the S&W inscribed to him as a token of remembrance. The above scenario has been passed down through word of mouth by the citizens of Shelbyville, Kentucky, for many years and is presumed to be true.

On January 24, 1863, Private Masonheimer was discharged at Murfreesboro, Tennessee, on Certificate of Disability by General Rosecrans. Heartbroken at this unhappy turn of events, Masonheimer returned to Shelbyville, Kentucky, and his occupation of merchant tailor.

Fortunately for him, Masonheimer had a further and heroic role to play in the North/South conflict. From the *Shelby Record Newspaper*:

> "On August 26, 1864, a raid was made by a band of Confederate guerillas on Shelbyville for the purpose of stealing a lot of guns that were stored in the Courthouse for

The courthouse in Shelbyville, Kentucky.

Town blockhouse on Main Street, Shelbyville, July 16, 1865.

the protection of its citizens. They came into town early one morning over the old dirt road, afterwards the Burk's Branch Pike. The confederate leader was Captain Dave Martin, who anticipated no trouble, but found out their mistake shortly after their arrival. Two prominent citizens who lived near this courthouse heard the noise of tramping horses and made an investigation and, discovering what was going on, began firing on the guerillas.

"These two brave men were Thomas C. McGrath, father of Will and James McGrath, of this city and a merchant tailor by the name of J.H. Masonheimer. Mr. McGrath did his shooting out of a third story window fronting the courthouse yard and Mr. Masonheimer stood in the doorway of his store on Main Street. A terrific battle was waged for sometime and the guerillas were finally driven off without the guns. Three of their men, Lt. Jo Veatch and two Privates by the name of Dale and Smith were killed and several of them wounded. Besides them, a Negro man named Owen, a blacksmith in the employment of James Hickman was also killed.

The brave men who so valiantly fought the Confederate guerillas narrowly escaped death or serious injury from the numerous bullets fired at them. One ball plowed through

> Mr. McGrath's forehead and scalp, making a painful but not serious wound, but Mr. Masonheimer escaped unhurt.
>
> The Board of Trustees of the town of Shelbyville, feeling that some official recognition of the bravery of the two men who drove the guerillas off made the following order at its next regular meeting held on September 22nd· 'It is ordered that a committee composed of T.B. Cochran, C.C. Watts and Gideon J. Stivers be appointed to ascertain in what manner this board can best compliment T.C. McGrath and J.H. Masonheimer for their gallant conduct in defending the town on August 26 against Confederate Capt. Dave Martin and his cut-throat gang.' The committee, at the next meeting of the Board, recommended the presentation of a pair of pistols and holsters to each of the gentlemen, estimating the cost at one hundred dollars. The report was approved and the committee was continued with instructions to purchase the pistols and see that they were handed over to Messrs. McGrath and Masonheimer."

At last, John Masonheimer had accomplished his heroic goal to defend his country and city against the rebels, even though he had mustered out of the service some nineteen months earlier. In effect, J.H. Masonheimer got a second chance at achieving heroic action!

Chapter Twelve

The Lefaucheaux Revolver of John Wysong, Civil War Soldier from Ohio

Collectable guns seem to find their way to interested buyers in many unusual ways. Such was the case with Lefaucheux revolver, Serial #26837 C1860, which was carried by Private John Wysong from the city of Pyrmont, in Montgomery County, Ohio.

I have a good friend, Dave Buehrle, whose girlfriend is Eloise Hendrickson. One day in 1993, Dave told me that Eloise had a Civil War revolver that she had inherited from her great-grandfather, John Wysong, but she was scared of guns and wanted it out of her house!

So, I called Eloise and invited her to lunch, asking her to bring the gun and any family photographs or history she might have. She responded that she would be happy to do this, but she needed time to get with two of her aunts in order to gather up any family history that existed; she would call me.

Impatiently I awaited her call, which seemed to take months. Finally, we got together and she presented me with a Model 1854, 12mm (about .45 cal.) double action Lefaucheux six shot, pinfire revolver in very good condition. It had a 6-1/4" round barrel, lanyard ring, ejector rod and loading gate. This was a very so-

John Wysong in rocking chair surrounded by his family members.

phisticated revolver in 1860, because it fired a large caliber cartridge and could be used in either single or double action mode. At that time, the most powerful American made cartridge revolver was the .32 cal. Smith & Wesson Old Army #2, which was only single action.

Eloise pleasantly surprised me with a folder full of terrific family history, including the following photographs:

a) John Wysong in a rocking chair with his wife, Emeline, seated and holding one of their grandchildren and surrounded by family members, ca. 1875. This front porch shot is of their Morgan Street home in Knightstown, Indiana.

b) Camp Dennison, northeast of Cincinnati, Ohio, where John Wysong was mustered into the army.

c) John and Emeline Wysong at their home in 1885.

d) John and Emeline Wysong in 1900. John was seventy years old in this photo.

In addition, Eloise gave me the family history going back to 1558, which included a newspaper article from the Fincastle, Virginia *Herald* that concerned the dedication of "Ludwig Wysong Memorial Park" right adjacent to Feidt Wysong's Blacksmith Shop at the corner of US 220 and Main Street in Fincastle, Virginia. The purpose of the park is to memorialize the fact that Ludwig Wysong was the father of five sons who were veterans of the American Revolutionary War. The Memorial reads as follows:

Dedicated in Memory of
LUDWIG WYSONG
1678 – 1784

And His Five Revolutionary War Sons

LEWIS	JOSEPH
1743 – 1808	
VALENTINE	FEIDT
1744 – 1824	1755 – 1837

JACOB
1757 – 1823
By Their Descendants

These descendants erected the memorial in 1978 during a four-day Wysong family reunion that attracted about 300 relatives from twenty-four states. The city of Fincastle, Virginia, agreed to make this memorial a dedicated park, which includes the adjacent blacksmith shop.

Camp Dennison near Cincinnati, 1861.

John and Emeline Wysong, ca. 1885

John and Emeline Wysong, ca. 1900

I had written this entire article in February 2004, thinking it was complete. Then, an amazing thing occurred. I was standing in line getting guest badges at the March 4, 2004, Ohio Gun Collectors meeting. I had been talking to the man in front of me and told him I had just written a magazine article about Civil War veteran John Wysong, and that I owned his Lefaucheux revolver. He smiled with a look of disbelief and introduced himself as Charles Wysong from Indiana, and explained that he had attended several Wysong family reunions and possessed large quantities of family history including his great-grandfather, John Wysong's Civil War sword and scabbard. Naturally, we agreed to exchange Wysong family information including photographs. In July 2004, I received a large manila envelope stuffed full of Wysong history including a picture of John Wysong's sword and scabbard, plus a complete family tree that measured about four feet square. This terrific additional information caused me to rewrite my article in November 2004, making it more interesting and historically accurate. Now I had the family tree from 1558 right down to Eloise Hendrickson, who had sold me the gun in 1993.

You can imagine my amazement at the fortunate coincidence of meeting Charles Wysong, a living family member and owner of John Wysong's sword and scabbard. After much research through the Wysong family records, it turned out that Charles' great-grandfather, John Wysong, served in the 12th Regiment of Indiana Cavalry and was the first cousin of my John Wysong, who served in the 71st Ohio Volunteer Infantry.

The family records show that Ludwig Wysong, born in Germany in 1678, was the great-great-grandfather and Valentine

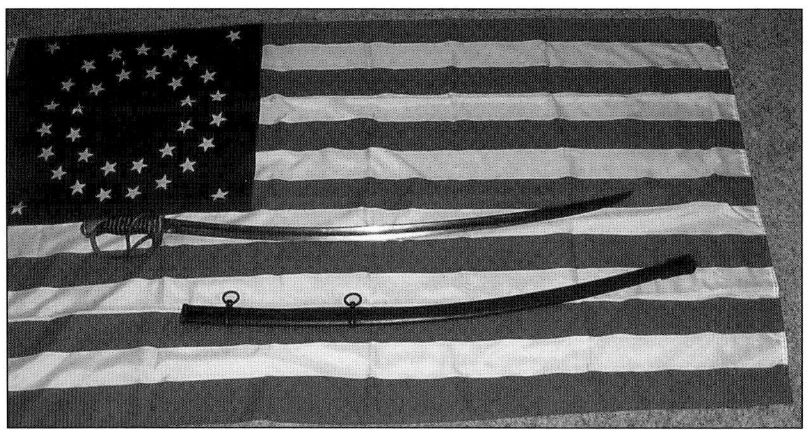

John Wysong's Sword and Scabbard

Wysong, Revolutionary War veteran born in York County, Pennsylvania in 1744, was the great-grandfather of our John Wysong, veteran of the Civil War.

Any Civil War gun collector is fortunate indeed to acquire a firearm that is not only identified to a particular soldier, but also includes his picture and extensive family background. I gladly purchased her Lefaucheux revolver along with a notarized affidavit of its history, and went about doing additional research including John Wysong's military service and pension records. At the time, I had little detailed information on the history of Lefaucheux revolvers and their role in our War of the Rebellion.

Fortunately, I had met Chris Curtis at the Nashville Civil War Show (Tennessee) in 2003. Chris specializes in Lefaucheux revolvers and in 2002 wrote a book about the Lefaucheux Model 1854, *Systeme Lefaucheux: Continuing the Study of Pinfire Cartridge Arms Including Their Role in the American Civil War.* Chris provided me with this very interesting information.

Following is a brief history of Lefaucheux type firearms in general and the Model 1854 revolver in particular:

Casimir Lefaucheux was born in 1802. He was evidently a prodigy in the field of mechanics, as records show he was apprenticed first in Sarthe, France, and later in Paris in the gun trade. He went to work at the Paris shops of Pauly, a famous gunmaker, at age 12. He continued there, becoming first manager and later buying the establishment in 1827. He invented the pinfire cartridge in 1835,

– Continued on page 129

Chapter One

Major Robert McClure's M 1862 Colt Police, .36 Cal. Revolver

Chapter Two

Captain Ruby's Colt M 1862 Police Revolver #3291

Inscription on backstrap of Captain James Ruby's Colt Police: Pres'd to J.N. Ruby by his Co. E 2nd Wis. Reg't Sept 21, 1861

Chapter Three

Colonel McNeill's Smith & Wesson complete with holster

Colonel McNeill's M1840 Cavalry Saber and Scabbard

Chapter Four

A matched pair of Smith & Wesson No. 2 old Army revolvers complete with holsters, .32 Cal. 6-inch barrels, Serial #s 7837 and 12905. This pair was carried by Second Lieutenant Samuel Cook, Co. H, 13th Indiana Volunteer Cavalry.

Smith & Wesson old Army No. 2 in a shooters box, .32 Cal., 6" barrel, Serial #29500. This gun was carried in the Civil War by Corporal Jacob Deisher, son-in-law of Lt. Samuel Cook, who was married to Sarah Cook, Samuel's only daughter. Corporal Deisher served in Co. D, 155th Illinois Infantry.

Chapter Five

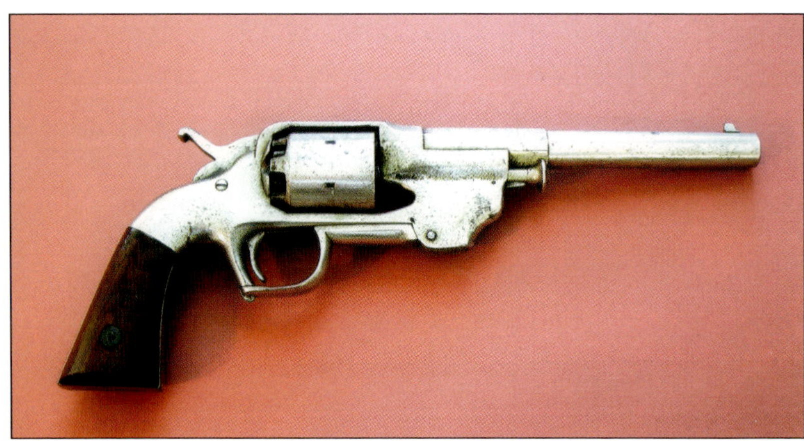

Allen & Wheelock, .44 Cal. Army, 7.5-inch barrel, Serial #88, issued to Private John E. Cranston, 3rd Michigan Volunteer Cavalry.

Lt. Thomas Brayton's Belgian made 7.65 mm (.30 Cal.), pinfire, cased with metallic cartridges.

Chapter Six

Sergeant Al D. Fobes Colt M1849 Pocket Pistol

The inscription reads, "Al D. Fobes Comp. E"

Chapter Seven

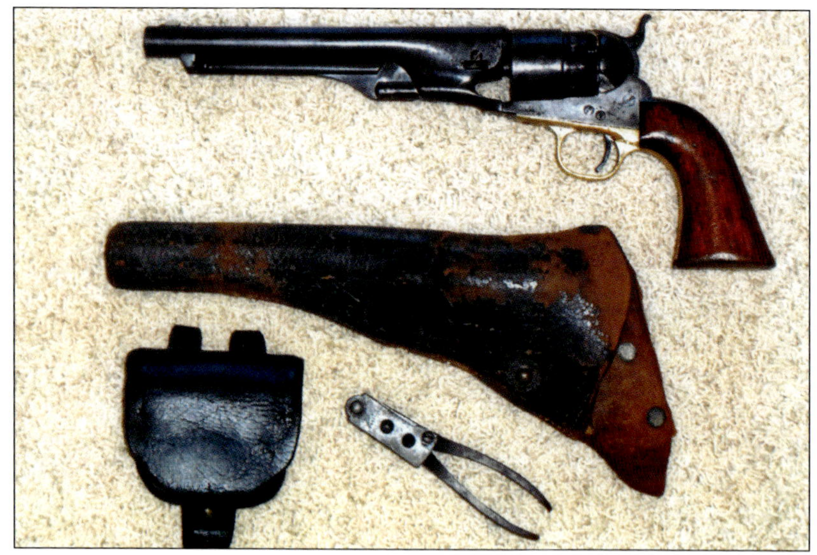

Private Charles Edward Coffey's Colt M1860 Army and Accessories

Colt M1860 Army, .44 Cal.

Chapter Eight

Private John Hancock's Bacon Revolver

The inscription reads, "John.Hancock.Peekskill, New York"

Chapter Nine

Colonel John Warner's Colt M1851 Navy, factory engraved with ivory grips

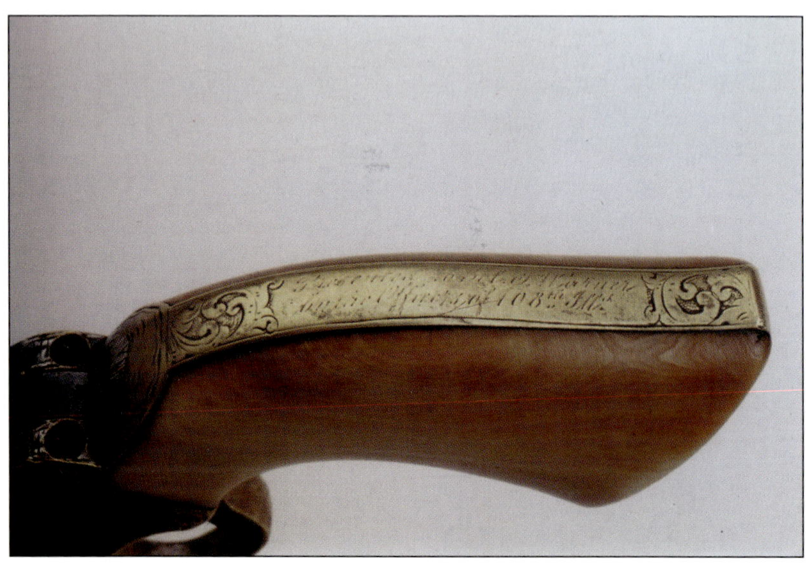

The inscription reads, "Presented to Col. John Warner by the officers of the 108th Illinois"

Chapter Ten

1st Lt. J.F.E. Chamberlain's Cased Smith & Wesson Old Army No. 2

The inscription reads, "Presented to J.F.E. Chamberlain by his friends at the U.S. Armory"

Chapter Eleven

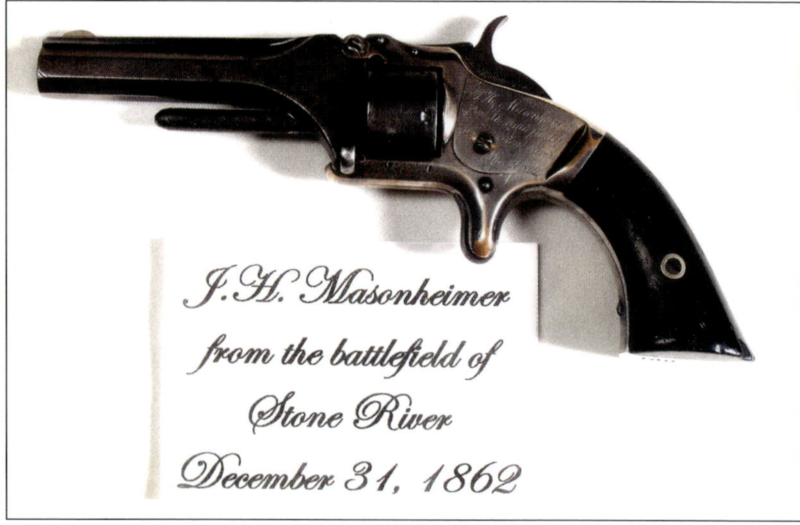

Private John H. Masonheimer's Smith & Wesson Model 1, Issue 2. The inscription is printed underneath the image.

Chapter Twelve

Private John Wysong's LeFaucheux Pinfire Revolver

Chapter Thirteen

Model 1878 Colt double action revolver, .45 caliber, 7.5-inch barrel, serial no. 14733, manufactured and delivered to the Canadian Department of Militia and Defence in 1885 to put down Louis Riel's Rebellion.

Chapter Fourteen

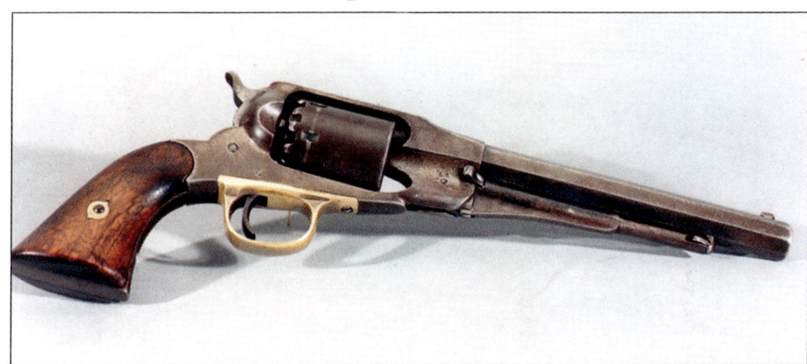

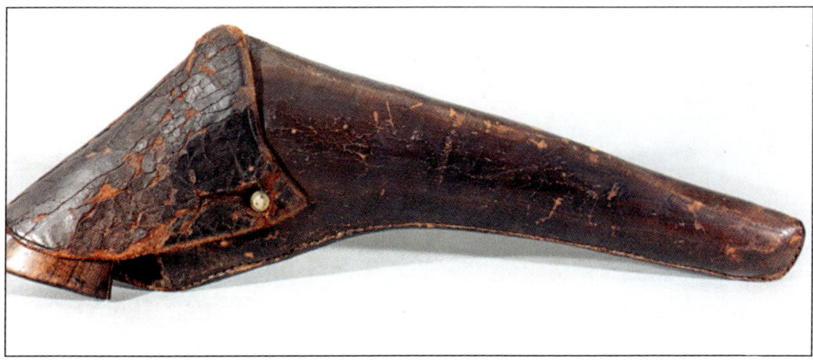

Private George App's Remington Navy Revolver and Holster

Chapter Fifteen

Sheriff John Bence Van Hagan's Colt M1849 Pocket Pistol

The inscription reads "J.B. Van Hagen" on the backstrap.

Serial No. 176985 with the inscription "Nevada City, Cala." on the butt.

Chapter Sixteen

Left side of the Smith & Wesson revolver belonging to John Baird, V.P. of the Metropolitan Elevated Railroad serving New York City.

Inscription on the backstrap reads "John Baird".

Chapter Seventeen

Judge James Kelly's Remington Ring Trigger Derringer inscribed "Judge James Kelley, Caldwell, Kansas."

Chapter Eighteen

Major General John C. Breckinridge's pair of Horse Pistols made by Kuckenreuter of Regensburg, Germany

Barrel showing gold inlaid master's mark and inlaid maker's name – J. Andreas Kuckenreuter

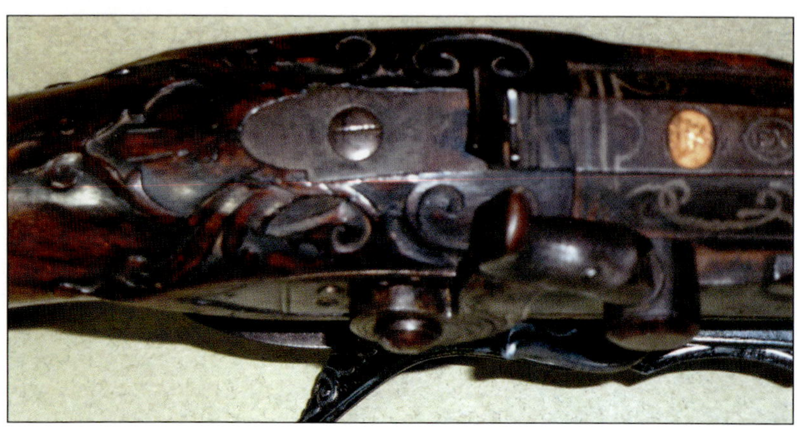

Handcarving at the breach

Chapter Nineteen

Cased Colt .41 Cal. #3 Thuer Derringer

The inscription reads "Wm. E. Bainbridge / Council Bluffs".

Chapter Twenty

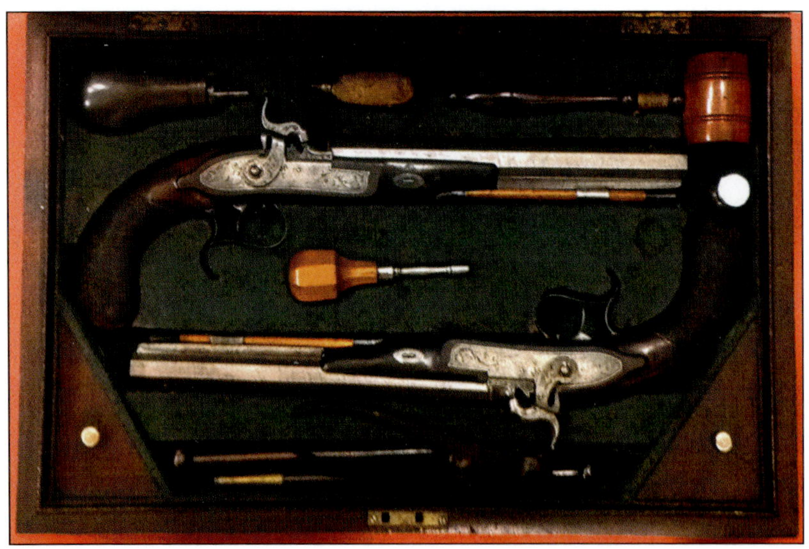

Cased set of pristine, untouched American dueling pistols with all original accessories by Prosper Vallee of Philadelphia.

Engraved lock showing the unusual duelist aiming pistol and P. Vallee's signature.

a fact obscured because it was a patent addition hidden within the actual patent for improvements to arms. He made firearms, both shotguns and single shot pistols, as well as underhammer pepper-boxes. They were all of high quality, and he won a Medal of Honor for his exhibit at the International Arms Show in London in 1851. He died the following year in Paris at the age of fifty. His only surviving son, Eugene, had evidently already inherited his father's genius for firearms design and had been exposed and apprenticed into that trade, also from an early age. Having attended the London Exhibition with his father, he was most impressed with the revolvers displayed there by Colt. Being only nineteen at the time of his father's death, he spent the next thirteen months studying in Belgium paying close attention to not only methods of manufacturing, but business dealings.

Eugene returned to Paris in 1853, assumed control of the family business, and began perfecting his idea of a large frame, military caliber revolver using his father's cartridge rather than the percussion system. He applied for patent rights on April 15, 1854, and as soon as the patent was granted in June, began limited production. He was careful to add his first initial "E" to the Lefaucheux markings to avoid confusion with his father's work.

Patented in France, Belgium, and England, the Model 1854 was an immediate success. Firearms evolution in the U.S. would have been quite different had he thought to extend his patent protection here. His patent predated the Rollin White patent by a full year. At the time of its introduction, the Lefaucheux Model 1854 revolver was the most advanced handgun in the world. Looking for a more modern revolver, the French Navy tested the Lefaucheux against many guns, including Colt and Adams, and adopted it for Naval use in 1857. Production began in 1858, and France was the first major power to adopt a cartridge handgun.

Under contract with the Navy, Eugene not only received a royalty but also was allowed to manufacture guns for private or foreign military contract sales. Spain adopted this revolver as well as Norway and Sweden. Italy also bought a large number. The U.S. Military looked into this revolver and actually tested a few in 1857 and 1858, but no action was taken. At the outbreak of the American Civil War, Colonel Schuyler was sent to Europe to purchase arms of different types for the Union. In October 1861, he purchased directly from Lefaucheux 10,000 revolvers with ammunition. In addition to that number, other U.S. dealers and importers sold about 2,000 more to the Federal government.

The accepted serial number range of the Federal guns has long been 25,000 to 37,000, but I have never been convinced that all those purchases were a single block of numbers. Research now seems to extend the range both up and down from this grouping. The Wysong revolver #26937 falls neatly into the accepted range even without any future readjustment of these numbers. It seems that most were issued to Western cavalry units, but we have seen them identified to New Jersey, Maine, New York, and others. No records have been discovered to show to what extent the Confederacy purchased this type, but some interesting records do show large amounts of 12mm ammo in a couple of Confederate Ordnance depots.

As the war progressed, logistical, ammunition, and other problems led to the Lefaucheux gradually being replaced by the older, but more familiar, percussion type revolvers. Still, it played a very important role in the War, being the fourth most widely issued revolver behind only Colt, Remington, and Starr. Total production of the Model 1854 reached about 120,000, and this number includes all styles, variations, and various foreign military contracts. Eugene manufactured many other types of pinfire cartridge arms and finally got into center fire late in his career, which ended in the early 1870s. Casimir and Eugene made firearms for 40 years, during which time they had registered over 160 patents, making them among the most prolific and important of firearms patent holders.

After reading over 100 pages of archive information from Washington, D.C. on John Wysong, I was able to develop a synopsis of Wysong's service and life in general as follows:

Private John Wysong
Co. I, 71st Ohio Volunteer Infantry

Born:	1830 in Montgomery County, Ohio, near Eaton
Complexion:	Light
Eyes:	Blue
Hair:	Light

Height: 5’ 11”
Weight: 135 lbs
Occupation: Brick Masonry Contractor
Died: February 17, 1907 at the age of 77
Mustered In: Enlisted February 28, 1861 at Pyrmont, Ohio, Mustered in December 20, 1861 at Camp Denison, Ohio, northeast of Cincinnati.
Mustered Out: Discharged at Columbus, Ohio on Certificate of Disability #292894 due to a knee wound, dysentery, irregular heart beat and lung disease. Mustered out officially at Huntsville, Alabama, February 10, 1865.
Married: Emeline (Harry) Wysong September 10, 1852, at Banton, Illinois, by Reverend Miner. Marriage license was issued from Lewistown, Illinois.
Children: John and Emeline Wysong had four children:

- Altha Isabella, born December 11, 1855
- Ambrose Gartland, born April 4, 1860
- William Edgar, born June 8, 1866
- Adali Daisy, born July 8, 1876

Residence: From 1830 to 1852, John Wysong lived with his parents, John and Rachel, on a farm near Eaton, Ohio. From 1852 to 1866, John and Emeline Wysong lived in Pyrmont, Ohio, then a few months in Indianapolis. From 1866 to 1907, John and Emeline lived on Morgan Street, Henry County, Knightstown, Indiana.
Occupation: In 1844, when John Wysong was fourteen years old, he apprenticed as a bricklayer for his first cousin, Jesse D. Wysong. Then John and Jesse became partners in brick masonry, contract-

ing until John enlisted in the Army in 1861. In 1866, after John Wysong spent about three years recovering from his war illnesses and injuries, he again became Jesse Wysong's partner in brick masonry contracting in Knightstown, Indiana. Jesse Wysong lived next door to John Wysong on Morgan Street.

Pension: Original declaration for a pension, #401,600, due to war injury and illness was filed January 28, 1880, with back payments to Wysong for $630 (about $40,000 in 2004 dollars). Wysong received $10 per month to November 11, 1885, $14 per month till February 6, 1889, $18 per month till February 1899, $24 per month till 1906, and $30 per month till his death on February 17, 1907. His widow, Emeline, continued to receive $30 per month until her death. Emeline inherited their paid for home on Morgan Street, which was valued at $1,200 in 1907 (a $1,200 home in 1907 equals about $100,000 today).

Battles:

Shiloh, Tennessee. . . . . April 6-7, 1862
Clarksville, Tennessee. . . . . August 19, 1862
Fort Donelson, Tennessee. . . . . August 25, 1862
Cumberland Iron Works, Tennessee. . . August 26, 1862
Clarksville, Tennessee. . . . . September 7, 1862
Jonesboro, Georgia. . . . August 31 to September 1, 1864
Lovejoy Station, Georgia. . . . . September 2-6, 1864
Columbia, Tennessee (Duck Run). . November 24-28, 1864
Nashville, Tennessee. . . . . December 15-16, 1864

It is interesting to ponder the magnitude of the damage the Civil War inflicted on our nation, not only the 600,000 lives lost, but also the endless suffering—financial, physical, and mental—

that remained with the living survivors. In the case of John Wysong, he was three years using a cane and recuperating from his wounded left knee, which occurred April 9, 1862, at "Bloody Shiloh".

Tombstone for John Wysong

General U.S. Grant recalled the carnage in his memoirs: "I saw an open field, in our possession the second day of battle over which the Confederates had made repeated charges the day before, so covered with dead that it would have been possible to walk across the clearing, in any direction, stepping on dead bodies without a foot touching the ground."

Shiloh was the greatest battle ever fought on American soil to that time. The Union sustained more than 13,000 casualties, the Confederacy more than 10,000, and John Wysong was there and was wounded! Armies paid a bitter price for carelessness and poor generalship on both sides.

In addition, he never recovered from his dysentery and heart/lung problems, which caused him to limit his manual labor to about thirty hours per week. By 1884, when Wysong was only fifty-four years old, he could no longer support his family. For the last five years of his life he needed continual personal care, all related to his wartime illnesses. He died February 17, 1907, and was buried at Glen Cove cemetery, Knightstown, Indiana. We all, in 2020, still owe a debt of gratitude to all the veterans who have given so much to preserve the wonderful America we have inherited!

Fortunately, soldier John Wysong, his story, and his revolver have not been forgotten! The Camp Dennison Civil War Museum in Cincinnati, Ohio, exhibited John Wysong's Lefaucheux revolver, his documents, and his story during the 2003 season for the benefit of thousands of enthusiastic visitors!

Chapter Thirteen

Colt Model 1878 Frontier Revolver Leads to the Notorious Louis Riel, A Unique and Misunderstood Canadian

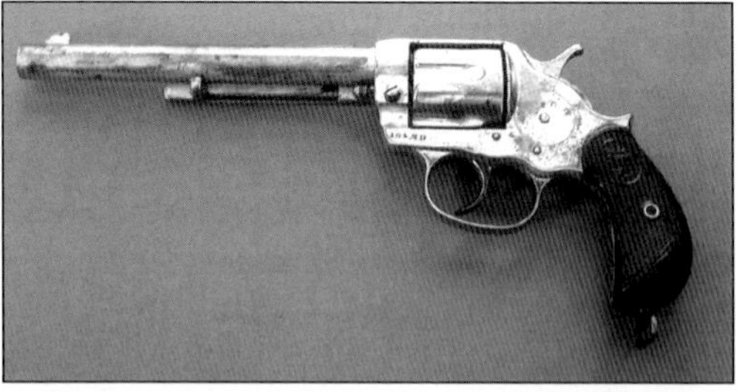

Model 1878 Colt double action revolver, .45 caliber, 7.5-inch barrel, serial no. 14733, manufactured and delivered to the Canadian Department of Militia and Defence in 1885. Bottom: Close-up view of M1878 Colt showing stamping 104.MD, indiating it was purchased by the Candadian Department of Militia and Defence.

Acclaimed writer, Mr. Charles G. Worman, author of *Gunsmoke and Saddleleather*, plus numerous excellent firearms books, has been the source of many identified guns in my collection.

In 1999, Chuck called me about an 1878 Colt marked 104.MD [See Fig. 1 and Fig. 2] indicating that it was purchased by the Canadian Department of Militia and Defence. The purpose was to arm Canadian troops against Louis Riel (ree EHL), a revolutionary who was at odds with the Federal Government of Canada in Ottawa, Ontario.

At the time, I knew nothing about the "Riel Rebellion," but the 1878 Colt with these seldom-seen markings appealed to me. So, I purchased it promptly. About the same time, the eminent author,

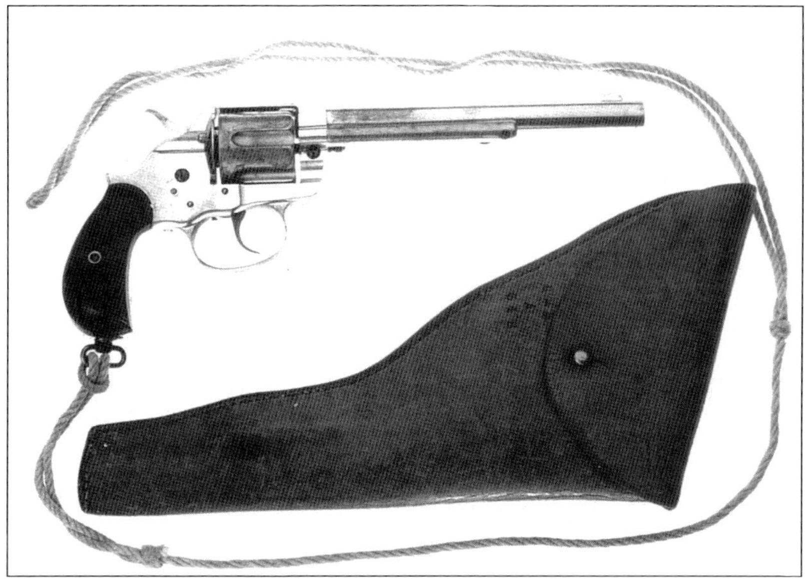

Serial No. 14618, .45 caliber, 7.5-inch barrel, nickel finish, rubber stocks. This revolver was one of a group of 120 revolvers sent to Hartley & Graham on April 14, 1885. It is marked 222.MD on the left side of the frame. The holster is an original militia defense issue, ca. 1885, with the marking "1-CMR-A-349" on the body of the holster. The CMR is a later stamp for Candadian Mounted Rifles. They fought in South Africa in 1900 during the Boer War. The holster is light brown leather with a full flap. Courtesy of Doug Carlson, Paul Goodwin Photograph

Mr. Don Wilkerson, published the book titled, *Colt's Double-Action Revolver – Model of 1878,* which, for the first time, laid out an in-depth historical study of the 1878 Colt Revolver, including nine pages of detail on the MD marked Colts. I read Mr. Wilkerson's book with great interest, and discovered the MD marked revolvers were quite an enigma, but definitely connected with "Riel's Rebellion."

The mysteries of the MD marked M-1878 Colts led me to research Mr. Louis Riel's life, which I thought might be rather obscure and difficult. Fortunately, World Book, Inc.'s Biographical Connections Division published, *Louis Riel with Profiles of Gabriel Dumont and Poundmaker,* in 2007. This excellent book brought together important details of Riel's Rebellion, but also laid out the entire history of Canada's national formation and its expansion from East to West, from the Atlantic to the Pacific, much like America's "Manifest Destiny." In addition, I found there were five earlier books on Riel, which assisted me in my research.

As an American history buff and a lifetime collector of handguns related to the U.S. westward development, I was somewhat shocked to learn how little I knew about Canada's parallel, although sixteen years later, westward movement and the prominent role American guns played in that epic Canadian story.

Per Mr. Wilkerson's study, in 1885, the Canadian Department of Militia and Defence (MD) purchased 1,001 Colt M-1878 Revolvers from Hartley and Graham Co., a New York City firm, for issue to the militia. The purpose was to put down The Canadian Northwest Rebellion, commonly known as, "Riel's Rebellion." The specifications for this revolver and accoutrement order were as follows: "Nickel-plated with hard rubber grips, .45 caliber, 7-1/2" barrels and lanyard rings." In addition, the Canadian government specified "Brown military-style flap holsters, cleaning and maintenance accoutrements, and thousands of rounds of .45 caliber central fire ammunition."

The result was that Canadian government troops made up of the Northwest Mounted Police (N.W.M.P. or 'Mounties') and the militia crushed the insurrection and hanged its leader, Louis Riel, for high treason on November 16, 1885.

Now, let's take a closer look at Louis Riel (1844-1885) and Canada's westward development through the following summary and chronology of Riel's life. Traitor or hero, madman or visionary leader, Louis Riel is one of the most controversial figures in Canadian history. Many French-speaking Canadians still see him as a hero, whose defiance of their federal government continues to inspire movements for cultural and even outright independence for Quebec. Other Canadians see Riel as a self-seeking rebel, who should have been hanged for high treason.

Louis Riel, 1844-1885

He is acknowledged as one of the founders of Manitoba; however, he also was regarded as a madman and was institutionalized for about two years. His own lawyers argued he should be found innocent of treason on grounds of insanity, but Riel himself claimed he was sane; it was the government of Canada, not he, that was insane.

A deeply religious Catholic, he left the Church because he believed that priests did not support the MĖTIS [mayTEE] cause. The MĖTIS cause was to get the federal government to extend them full democratic rights and to recognize MĖTIS land ownership, even though they had no deeds, but held the land by custom for generations.

MĖTIS are people of mixed European and Indian blood. MĖTIS, meaning "MIXED," could include European blood of any nation combined with Indian blood of any tribe, but the majority in Canada were of French and Indian blood. Riel was himself a MĖTIS of French and Indian origin.

He was brought up speaking the French language and practicing the Roman Catholic faith. His family had strong ties to Quebec, even though he was born in Manitoba. This French,

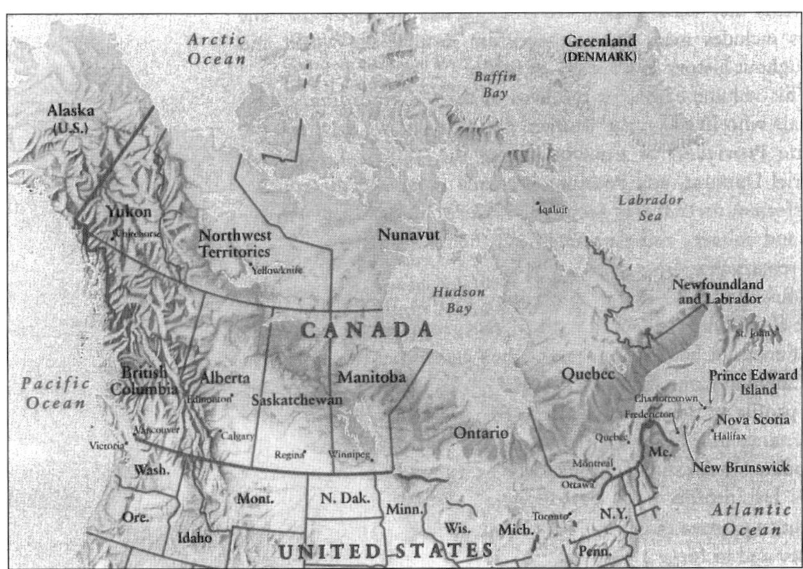

Louis Riel, Gabriel Dumont, and Poundmaker lived on the western plains of Canada during the 1800s in what we now call the Prairie Provinces: Alberta, Saskatchewan, and Manitoba.

Roman Catholic identity made him a hero among French Canadians, while making him an enemy to many English speaking Protestants. His life's work and its interpretation have shaken Canada since his death and continue to affect Canadian attitudes to this day.

Louis Riel was born October 22, 1844, in the Red River settlement, now St. Boniface, Manitoba. He was baptized at the Cathedral of St. Boniface, the Riel's parish church, in what is now Winnipeg. The Riel family were close, kindhearted, and strict Roman Catholics. The Riel family was also prosperous and enjoyed social status among the MÈTIS community. Louis Riel grew up free from the typical hardships of frontier life.

Riel was successful in school and attracted the attention of Bishop Alexander Taché, who arranged for young Louis to travel 800 miles east to Montreal, Quebec, in 1858, where he entered the Catholic seminary to study for the priesthood.

Louis found the Catholic seminary to be extremely disciplined and austere, while Louis was charming, outgoing, and had interest in both a wife and children. Louis left the seminary in 1865, and soon found employment in Montreal as an understudy to an outstanding lawyer, Rodolphe Lalamme. However, Louis found the work uninteresting, but stayed on for a year. He fell in love for the first time with Marie Guernon, but when her parents discovered the romance they broke them up because they did not consider a MÈTIS to be a suitable husband for their pure French-blooded daughter.

Riel was bitter, left Montreal, and worked odd jobs in Chicago and St. Paul, Minnesota. During this time, on July 1, 1867, Canada went through a momentous change. Canada became a dominion of the British Empire, a united self-governing nation. The new nation had a national parliament based on the British system, with Ottawa, Ontario, as its capitol.

The *Toronto Globe* newspaper lobbied for a transcontinental railroad so that Canada could expand across the western plains. Now the provinces of Manitoba, Saskatchewan, and Alberta could hook up with British Columbia on the west coast. All these changes made Louis Riel feel the hunting, fishing, and gathering lifestyle — practiced by the MÈTIS and Indians of the western plains— would be threatened, and the MÈTIS lands that they held by custom known now as 'Squatters Rights,' could be lost to the English system of surveys and deeds.

By 1869, this westward expansion, an influx of English-speaking people from Ontario, called Canadians, brought striking social and cultural changes within Louis Riel's Red River Settlement. Some of the newcomers were anti-Catholic and anti-French. Their intolerance strained relations with the resident MÉTIS people and Indians. The federal government in Ontario sent English Protestant surveyors into the Red River Settlement and eventually all of Manitoba. This deepened widespread MÉTIS insecurity about land ownership. The simultaneous happenings of a Dawson Road building crew and the Federal surveying team were to become the catalyst for rebellion. From this open rebellion emerged a talented new MÉTIS leader Louis Riel, who was only twenty-five years old.

Riel was passionate about the MÉTIS cause, a gripping speaker, charming and well educated, and spoke fluent French and English. In October 1869, Riel led a party of MÉTIS to repel the survey team, which walked off the job when threatened.

Louis then formed a national committee of MÉTIS and forbade the federally appointed leader, Governor William McDou-

The Canadian Pacific Railway, completed in 1885, linked British Columbia's Pacific Coast region and eastern Canada and opened the country's expanse of plains to intensive agriculture and development. With the advent of the railroad came many settlers to Saskatchewan.

In December 1869, Riel proclaimed the establishment of a provisional government based largely upon the National Committee of the Metis. Riel, center, is shown surrounded by his associates in 1869.

gall, to enter the territory without the committee's permission. He then led the MÈTIS militia with his close friend, Gabriel Dumont as Commander, to capture Fort Gary, now Winnipeg, Manitoba. Louis Riel and his supporters proclaimed a provisional government, and Riel formally took over as president.

In 1870, the provisional government executed Fort Gary prisoner, Thomas Scott, a Protestant militant. This would prove to be Riel's greatest political mistake, an act that outraged the English-speaking Canadians. So, the Federal government under Prime Minister John McDonald, threatened to send a Federal force to squash the revolt. Fortunately, cooler heads prevailed, and the provisional government composed a list of proposed rights for the MÈTIS and chose a delegation under N.J. Ritchot, a Roman Catholic priest, to go to Ottawa and negotiate a settlement.

Ritchot was successful with all points of negotiation, but one. He was not able to obtain a written amnesty for Riel and his associates, although he was given verbal promises of amnesty. Instead, McDonald dispatched an army to the Red River area, declared Louis Riel an outlaw and passed the Manitoba Act, creating the Canadian Province of Manitoba as proposed and named by Louis Riel.

With no written amnesty, Riel, fearing for his life, fled to the United States and settled in the MĖTIS community of St. Joseph, Dakota Territory. Riel proclaimed he was satisfied, saying, "No matter what happens now, the 'Rights of the MĖTIS' are assured by this Manitoba Act. It is what I wanted; my mission is finished."

Over the next four years, ever since Louis Riel fled in late 1870, MĖTIS friends and associates urged him to run for a seat in the Canadian parliament. No one knew how Riel would be received in Ottawa should he be elected to the House of Commons. Nevertheless, he decided to run for election. Although Riel never set foot in Manitoba during the campaign, he won the seat easily in February 1874. Rumors swept through Ottawa that Riel was coming to claim his seat. The authorities sent platoons of police to search for Riel in the Parliament buildings and grounds, but did not catch him.

Riel arrived in Ottawa and somehow managed to sign in on the member's roster, proving that he was there. Notwithstanding his victory, Riel felt he was a marked man and fled Canada once again and continued the life of a fugitive.

The politicians in Ottawa finally pardoned Riel in 1875, but only on the condition that he be banished from Canada for five years. He accepted these onerous terms, but felt despair. He had tried many times to serve his country, but was always denied. Near the breaking point, Riel felt a lack of direction and purpose in his life.

Demoralized and mentally and physically fragile, Riel's intense religious devotion turned into megalomania (a mental disorder with delusions of power and importance). Riel claimed to have had an angelic vision where he was told, "Rise up, Louis David Riel, you have a mission to perform."

In July of 1875, Bishop Ignace Bourget of Montreal wrote Riel, "God, who has guided and assisted you has given you a mission that you must in all respects fulfill." Riel took this letter as a divine mandate to lead and protect the MĖTIS people of Canada.

This continued megalomaniac behavior led physicians to recommend that Riel be committed to an insane asylum in Longue

Pointe, Quebec. Riel was admitted under the assumed name of Louis R. David in March 1876. After two years, physicians released Riel from the asylum, advising him to lead a "quiet life."

Riel then drifted through the northeastern United States, where he met and fell in love for the second time with Evelina Barnabè of Keesville, New York. Riel, seemingly cured of mental disorders and anticipating marriage, began traveling widely in the United States in an attempt to find a job that would support a family. During long absences, Evelina and Riel exchanged letters expressing their devotion to each other.

In October of 1878, Riel concluded that his destiny would be in the American West, rather than the eastern United States. Evelina was not at all happy about moving west, where Indian wars were still going on in the 1870s. Riel would not go east and Evelina would not go west, so they broke off their engagement and never saw each other again! This was the second love of his life and now, it too, was lost.

By 1881, Louis Riel had resettled in St. Joseph, Dakota Territory, because it was close to Manitoba. Friends and family could visit and bring news from Canada. Riel learned a land boom was on and white English-speaking settlers were now dominating Manitoba Province.

Riel decided to locate further west in Montana territory, because it was peopled with small MÈTIS communities that were comfortable for him. He did odd jobs, sometimes serving as an interpreter between MÈTIS and U.S. Government officials.

Later that year, Riel settled in Carroll, Montana, and married a young MÈTIS woman, Marguerite Monet, with whom he had a son, Jean, in 1882 and a daughter, Marie Angelique, about a year later.

Louis Riel became a U.S. Citizen in 1883, but for years had continued to concern himself with the well being of the MÈTIS people. Typical of Riel's behavior, he saw U.S. citizenship as a personal aggrandizement, describing his naturalization as a "happening" that erased the border between the Canadian Northwest and the United States. In mid-1883, Riel, no longer banned from Canada, made an extended visit to Manitoba, staying at his Mother's home in St. Vital.

Later in 1883, Riel accepted a teaching job in Montana at St. Peter's Mission, which was run by Roman Catholic Jesuit priests. The routine of teaching did not suit Riel, because he really yearned for an important role on the stage of life. Riel was, at this point, a politician for better or for worse.

In 1884, a delegation of four MĖTIS from the Saskatchewan River Valley, west of Manitoba, visited Riel in Montana. Leader of the delegation, Gabriel Dumont, was a respected buffalo hunter and trader, known for his excellent reputation.

The message this delegation brought to Riel was that the MĖTIS of Saskatchewan had concluded they were running out of options in protecting their rights and their land holdings. Further, they concluded that Louis Riel was the right and *only* man that could help them! By 1884, Riel had achieved mythical status among most MĖTIS peoples, and only he could do for Saskatchewan what he had accomplished for Manitoba in 1870.

It is true that in many ways the situation in Saskatchewan in 1884 was similar to Manitoba in 1870, but there was one crucial difference — the rapidly expanding Canadian Pacific Railway was bringing endless settlers to the Canadian plains, including Manitoba and what is now the provinces of Saskatchewan and Alberta. The many settlers, and the railroad with its hunting parties, were contributing to the disappearance of the buffalo herds. The Indians and the MĖTIS of these vast plains were dependent upon the buffalo for their way of life. Many were beginning to starve, both from the dwindling buffalo and crop failures caused by drought. Petitions for help were sent to Ottawa, but to no avail.

The day after the delegation had invited Riel to return to Saskatchewan and lead the fight, Louis decided he would fulfill his "Divine Mission." So, he gathered his family and possessions to begin the 700-mile journey northward in late June 1884. It appears that this unbelievably strong motivation to put himself and his family in jeopardy could only come from a combination of a "divine duty to the Canadian MĖTIS cause" and "megalomania", which made Louis feel all-powerful. His mind was fragile.

By late July 1884, Riel, his wife, Marguerite, and their two children had arrived in Batoche [buhTAHSH] in Saskatchewan territory. Batoche was the center of MĖTIS culture, nestled on the east side of the Saskatchewan River. Batoche would be their home for about a year and with no funds, they depended on charity from friends, relatives, and the various MĖTIS communities of the region. Meanwhile, Louis Riel, tribal leaders, MĖTIS activists, and white farmers went about the complex process of trying to combine and present a unified front to deal with the Canadian federal government in Ottawa.

Finally, in December 1884, a petition crafted by Riel and his associates was dispatched to Prime Minister John McDonald's cabinet in Ottawa. In brief, the petition asked for the following:

Advocated confirming land titles to all land currently occupied by MĖTIS people.

Called for economic aid and increased government help for all Indian tribes of the area.

For oppressed white farmers, there was a recommendation for construction of a railroad north to Hudson Bay so they could ship their harvests to market.

Saskatchewan was to be granted provincial status so that its residents would have representation and participation in the federal government.

The prime minister and cabinet in Ottawa acknowledged receipt of the petition. Now that the petition was finished, Riel wondered if he should return to Montana, since he had no means of support, except charity.

In 1885, Riel's supporters, though dwindling in number, convinced him to remain in Saskatchewan to await an answer from Ottawa — unfortunately, none ever came.

During this uneasy winter waiting period, Louis Riel sometimes became extremely agitated and fell into fits of ranting, occasionally threatening MĖTIS violence against everyone else. He began to hint of a "war of extermination" against all those against "Our Rights." In his wild rantings, he advocated secession from the Church in Rome and the formation of a new church with a new Pope in the Canadian West. Many potential constituents saw evidence of mental instability, but to Riel's faithful followers, he remained their visionary leader.

With no answer from Ottawa, Riel was the first to utter the unthinkable to his inner circle, the prospect of taking up arms. The

Poundmaker, 1842-1886

Gabriel Dumont, 1837-1906

Catholic clergy, frightened by Riel, condemned armed resistance and threatened to excommunicate anyone who participated in it.

On March 19, 1885, Riel proclaimed a provisional government with Pierre Parenteau as President. Riel was never officially part of the provisional government, but was its unofficial leader.

Prime Minister John McDonald and his cabinet considered the provisional government a deliberate provocation. They reacted fast, sending Major General Frederick Middleton, with about a thousand troops, westward by rail to reinforce N.W.M.P. forces already in the Northwest region.

The first battle took place at Duck Creek on March 26th, with Gabriel Dumont faced off against the Federal troops. Dumont's MĖTIS forces were well positioned defensively and defeated disorganized government forces under Officer Leif Crozier of the H.W.M.P.

Meanwhile, Indian violence under Cree Chief Poundmaker broke out. The village of Battleford was ransacked and burned. Farther north at Frog Lake, Chief Big Bear's warriors massacred a number of white British settlers. Many tribes were not specifically allied with Riel's forces, but they did share a common enemy—Federal forces. Some chiefs were being told Riel had supernatural powers in order to garner more united Indian support.

By April, part of General Middleton's army numbering about 800 soldiers was advancing toward Batoche. In an attempt to fend off Middleton, Dumont and Riel planned a surprise party in a gulch near Fish Creek. Dumont again positioned his men wisely and drew Middleton's forces into the gulch, where Dumont's men could massacre the enemy from higher ground with better cover. The result, on April 24, 1885, about 150 MÈTIS defeated 800 Federals, who retreated with heavy losses. Nevertheless, as each day passed, the Federals were gaining power because the Canadian Pacific Railroad continued delivering more men and fresh supplies from the East.

Indian problems continued. Lieutenant-Colonel William Otter with 300 soldiers, two cannons and a Gatling gun advanced on Chief Poundmaker's "Cut Knife Creek Reserve." Otter intended to punish Poundmaker for his raid and burning of Battleford a month earlier.

Poundmaker's warriors had only enough time to send the women and children into hiding before the Federals attacked. A seven-hour battle erupted on May 2nd, during which the tough, skillful warriors inflicted heavy damage on the Federals. Lieutenant-Colonel Otter had enough bruising, so he retreated, giving the day to Chief Poundmaker. After this, Poundmaker and his warriors rushed to support Riel at Batoche, where a final showdown was imminent.

Meanwhile, General Middleton consolidated his army of over 1,000 men and re-supplied them with cannon, ammunition, and plenty of food and water. The siege of Batoche, the MÈTIS capitol and stronghold, was about to begin. In addition, Middleton sent an armed steamboat, the *Northcote*, to attack Batoche from its vulnerable riverfront exposure.

Fortunately, Dumont's men were aware of the *Northcote* and disabled it with cables, which tore off the top floors of the steamer before it could reach Batoche on the Saskatchewan River. May 2nd, the siege began with only about 200 MÈTIS inside the fort to defend it. Poundmaker's warriors had not yet arrived. Dumont's men were well positioned and well trained. They fought valiantly for three days and nights till the MÈTIS forces ran out of ammunition and fled to the woods.

General Middleton was now in control of Batoche. Dumont's men finally fled south into the United States, but Louis Riel voluntarily gave himself up on May 15, 1885. Poundmaker and his warriors surrendered on May 26, 1885.

Ottawa wanted Riel tried for high treason, punishable by death, in an unfriendly venue. So, they transferred him far south to Regina where there was no MÈTIS support or sympathy for the man or his cause. Riel was jailed with ball and chain for sixty-five days until the trial commenced on July 28, 1885. Six jurors were selected — all were English-speaking Protestants; none were French Canadian, MÈTIS, or Catholic.

The court was packed to see the "Mad Rebel." The Judge, Hugh Richardson, spoke no French, yet most of the testimony would be in French. Judge and jury depended on translators. The Prosecutor's case depended on Riel's orders to carry out an armed rebellion and his wild threats to lead a 'war of extermination.' Riel was also accused of inciting an Indian War against the white settlers.

In July 1885, Riel (standing in the prisoner's box) was tried for high treason in a Regina courtroom. All of the jurors were English-speaking Protestants. Many observer thought that Riel's impassioned but rational speech may have discredited the insanity plea and sealed his doom.

The defense team relied on witnesses to bolster the claim of innocence by "reason of insanity." Riel was deeply troubled by the claim that he was insane. In open court, Riel testified:

> "I cannot abandon my dignity. Here I have to defend myself against the accusation of high

treason, or I have to consent to the animal life of an asylum. I don't much care about animal life if I am not allowed to carry with it the moral existence of an intellectual being.

"...I worked to better the conditions of the people of the Saskatchewan at the risk of my life...Manitoba, when the Government at Ottawa was not willing to inaugurate it at the proper time. I have worked till the inauguration should take place, and that is why I have been banished for five years...

"Even if I was going to be sentenced by you, gentlemen of the jury, I have the satisfaction if I die — that if I die I will not be reputed by all men as insane, as a lunatic...

"If you take the plea of the defence that I am not responsible for my acts, acquit me completely since I have been quarreling with an insane and irresponsible Government. If you pronounce in favor of the Crown, which contends that I am responsible, acquit me all the same. You are perfectly justified in declaring that having my reason and sound mind, I have acted reasonably and in self-defence, while the Government, my accuser, being irresponsible, and consequently insane, cannot but have acted wrong, and if high treason there is it must be on its side and not my part."

Most observers thought Riel's testimony was rational and well thought out, so his lawyer's insanity plea was discredited and Riel sealed his own doom.

The jury deliberated only an hour, returned a verdict of guilty, but recommended mercy in sentencing. Judge Richardson was under heavy pressure to impose the death penalty, because the politicians in Ottawa wanted Riel out of their way. The sentence was "Death By Hanging" to be carried out on September 18th in Regina, Saskatchewan Territory.

Meanwhile, the federal government was imposing harsh justice on scores of westerners involved in the uprising. Eight Indians were hanged in Battleford, and dozens of other MÉTIS and Indians were imprisoned.

On September 17th, one day before Riel was to hang, his case was appealed to Manitoba and the Privy Council in London, both of which upheld the sentence, but rescheduled the hanging for November l6, 1885. On that chilly, dark morning, Riel was attended by Father André, and spoke: "I thank God for having given me the strength to die well. I am on the threshold of eternity and I do not want to turn back. I die at peace with God and man, and I thank all those who have helped me in my misfortunes."

On the scaffold, Riel forgave all his enemies and asked their forgiveness in return. The executioner hooded Riel's head and sprang the trap; Riel died almost instantly!

On December 12, 1885, Riel's body was buried in the cemetery of the Cathedral of St. Boniface, Manitoba, where he had been baptized in 1844. In attendance were all his family members and many others who had known and loved him. Today, St. Boniface Cathedral is in the city limits of Winnipeg, Manitoba. A final unfortunate footnote: his wife, Marguerite, died in May 1886, daughter, Marie, died of diphtheria in her teens and son, Jean, died in his twenties in an accident.

Within twenty years, the Canadian West was transformed beyond all recognition. Saskatchewan and Alberta became part of the Canadian Confederation as provinces in 1905. The Canadian prairies now became one of the largest grain producing regions in the world. Later, they would find billions of barrels of oil and gas and support populations of millions. The Canadian West had no room for the MÉTIS lifestyle from an earlier age.

Riel's legacy still rattles Canada to this day! French separatists gained power in Quebec from the 1960s through the 1990s. Then there was a Provincial Referendum in October 1995, which came very close to separating this province from the rest of Canada.

Since about 1960, many Canadians have re-evaluated the contributions Louis Riel's life and work have made to the development and westward expansion of Canada. He has been formally recognized for his role in founding and naming Manitoba. In 1985, the federal government published a five-volume bilingual edition of Riel's works. In a 2005 ceremony to commemorate the 120th anniversary of Riel's death, a procession was led from St. Boniface Basilica to the rebel's gravesite in Winnipeg. The people were carrying flags with Louis Riel's name imprinted. The issue of 'hero' or 'traitor' is still being debated.

In Louis Riel's interview for the *Winnipeg Daily Sun* newspaper in 1883, he stated, "If the people of Canada only knew the grounds on which we acted and the circumstances under which we were, they would be most forward in acknowledging that I was right in the course we took, and I have always believed that as I have acted honestly, the time will come when the people of Canada will see and acknowledge it." The tragedy and enigma is that this recognition came over 100 years too late!

In a 2005 ceremony to commemorate the 120th anniversary of Riel's death, a procession was led from St. Boniface Basilica to the rebel's gravesite in Winnipeg.

Having laid out Louis Riel's life and Canada's western expansion, we can now return to the enigma of the 1,001 Model 1878 Colt revolvers that were sent to the Canadian Federal Government in 1885 to help stamp out the "Riel Rebellion." According to Mr. Wilkerson's book, *Colt's Double Action Revolver - Model of 1878*, probably only 200-300 of these Colts actually were issued to the Canadian troops, both militia and N.W.M.P. during the uprising. Another 200-300 were issued to Canadian militia shortly after the 1885 rebellion. The balance, around 500-600 revolvers, remained in storage with the Canadian Department of Militia and Defence until the 1900 Boer War in South Africa. The war involved the British Empire, so several

Canadian units were enlisted to help fight and win. The Canadian 1st Contingent, Special Service Force was armed with about 300 Model 1878 Colt revolvers left over from the "Riel Rebellion" in 1885. The 2nd Contingent, N.W.M.P. and Western Cowboys, were armed with Colt's New Service Revolvers.

The remaining 300 M-1878 Revolvers in storage with the Militia and Defence Department were issued to Canadian Forces in World War I. Featured in Don Wilkerson's book on page 256 is a photo of a member of the 63rd Halifax Rifles sporting a brown flap holster with flap open to show a nickel-plated Model 1878 Revolver.

Mr. Wilkerson concludes that only about 300 of the 1,001 revolvers ever actually received the MD markings. Quoting Mr. Wilkerson, "It is not known with any degree of certainty just when and by whom the marks were applied."

The above information illustrates that there is a great deal of speculation, creating an enigmatic aura around the exact usage of Canada's 1,001 M-1878 Revolvers.

In conclusion, we have the incredible life story of Louis Riel, an enigmatic and important historical figure paired up with an enigmatic and mysterious shipment of 1,001 Model 1878 Colt Frontier Revolvers, designated to help destroy "Riel's Rebellion" and Louis Riel himself.

Chapter Fourteen

Mule Skinner George App, His Remington Revolver, and the Western Indian Wars

The Remington revolver serial #36622 and original holster pictured here was carried by Private George App, Company D, 3rd U.S. Infantry, from 1873-78 in the Western Indian Wars. He enlisted November 6, 1873, at New York City, and was described as being 22 years old, 5'9" tall, and 140 lbs., with gray eyes, light hair, and fair complexion.

Private George App was assigned to duty as a mail wagon driver. The nickname for App's grueling Army job was "Mule Skinner" because, rather than riding on a seat on the mail wagon itself, the mule skinner would ride on one of the wheel mules and crack the whip on the four mule team, hence "Mule Skinner." According to family history, App acquired the revolver and holster for his personal use, it being much less cumbersome than a rifle

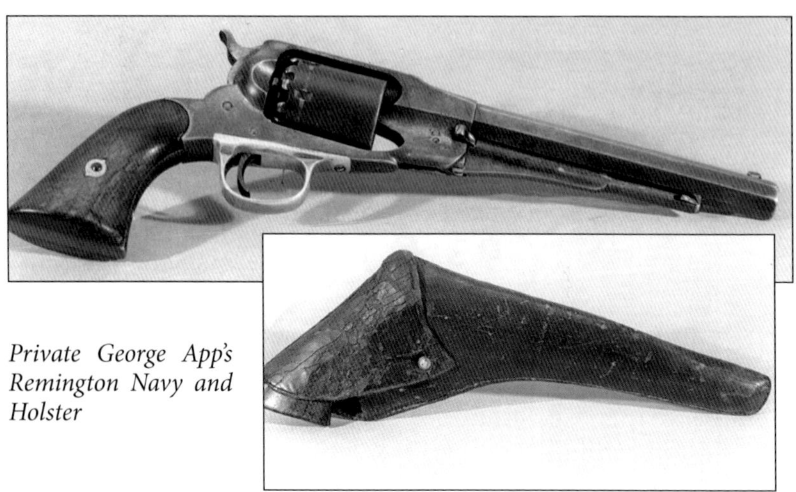

Private George App's Remington Navy and Holster

while engaged in such duty. I'm sure, in his youth, Private App looked forward to the great and invigorating adventure before him.

Years later, the following newspaper clipping from the *Bridgeport Connecticut Post*, July 27, 1933, told the history of what actually happened through an interview with George App. In 1933, he was 82 years of age and would die about three years later on May 20, 1936:

> "He began active service in Indian Territory, 1873; being almost continuously on the march and patrolling. App drove the mail wagon, Camp Supply to Fort Dodge, Kansas — 'soldiers kept the mail routes open.' He helped defend the mail wagon in an Indian attack, June 19, 1874 on Buffalo Creek; recalled the death of a white buffalo hunter who traveled alone — 'The hunter was mounted on a fast horse, and contended he would be safe enough alone for his pony could out run any horse the Indians owned.' He was missing one night and the next day a search party of soldiers was attracted by turkey buzzards circling over his dead body. He paid the price of traveling alone and underestimated the prowess of the Indian scalp hunters. Orders were to save the last bullet for yourself when fighting the Indians.
>
> "Mr. App, born in 1851, enlisted in the Army in 1873, serving first at Camp Supply, Indian Territory, then on to

New Orleans for duty during the great railroad strike of 1877, and then to duty stations in Montana until discharge late in 1878. The soldiers made their own roads on their trip through the western Montana wilderness and across the mountains. Crossing the Great Divide, they could find no water, and had to melt snow to make coffee. At one time a blizzard suddenly set in and they were snow bound for five days. They had tents for shelter, but the cold was intense and there was a careful check each morning to see if any succumbed to the cold over night. Several of the soldiers died in their sleep and were found frozen stiff the next morning.

"Indians in the West were often better armed than were the Army soldiers. Traders who looked only for profit, readily equipped the Indians with the best rifles of the times in return for furs and buffalo robes which were in great demand in the East.

"Private App's five-year enlistment was ended late in 1878. He did not re-enlist, for the Army at that time offered no career for ambitious young men."

Private App's Remington revolver has excellent provenance coming from the collection of the late Mr. Don Rickey, Jr. noted Western historian, former Superintendent of the Custer Battlefield National Military Park, and author of several books including: *Forty Miles A Day On Beans And Hay*, a history of the U.S. Cavalry during the Indian Wars. Mr. Rickey acquired this revolver directly from George App's daughter, Mrs. Catherine App Albrecht of Garfield, New Jersey, in October 1954. Fortunately, she also loaned Mr. Rickey additional relics: a photograph of Private App and two other soldiers in Infantry dress uniforms, a pair of brass Army spurs, and a braided leather whip App used as a mail wagon driver.

I acquired the gun and holster in 1994 from Mr. Charles G. Worman, a noted historian.

Summarizing the chain of ownership goes as follows:

1873-1936 Private George App
1936-1954 Mrs. Catherine App Albrecht
1954-1993 Mr. Don Rickey, Jr.

1993-1994 Mr. Charles G. Worman

1994-2013 Mr. Jerry W. Pitstick

In 140 years, this gun has had only five owners. Detailed information on percussion Remington revolvers is much less common than for Colts, for example, so this information is fortuitous indeed.

In summary, the life of a U.S. Army soldier during the Indian war era was dangerous, arduous, low paying, and with very limited upward potential, not to mention, the very limited supply of female companionship. For the Indian, it was tragic, hopeless genocide, chaos, starvation, lost land, and the end of a culture.

It seems only we people of the Third Millennium can look back, with unbridled nostalgia, to the taming of the West, and consider the experience exciting and dramatic.

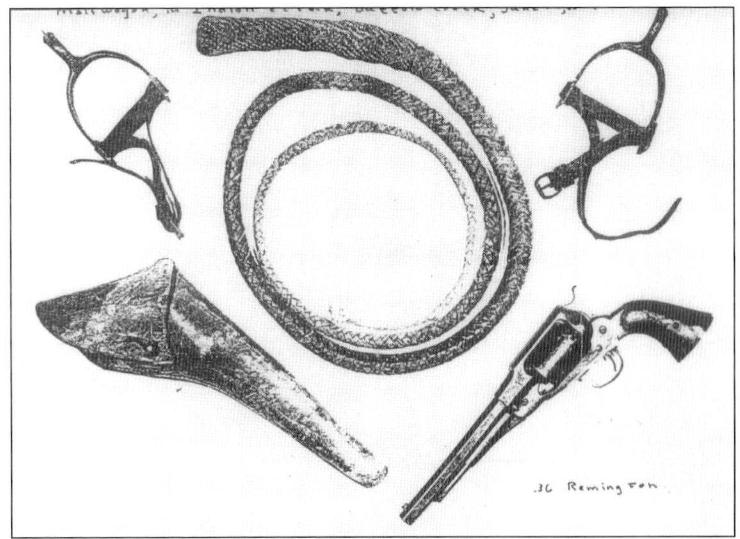

.36 caliber Remington and accoutrements belonged to Private George App

Chapter Fifteen

John Bence Van Hagen – Early California Sheriff and His Inscribed Colt M1849 Pocket Model

I purchased this Colt '49 pocket model in 1984 from Charles G. Worman, and also purchased a copy of his book, *Firearms of The American West 1803-1865,* because this revolver is pictured on page 279.

As shown, the gun is inscribed "J.B. Van Hagen, Nevada City, CALA". In his book, Worman indicates some of its history contained in his letter of transmittal to me.

The era of California history from the discovery of gold to the Civil War is one of the most colorful in America's history,

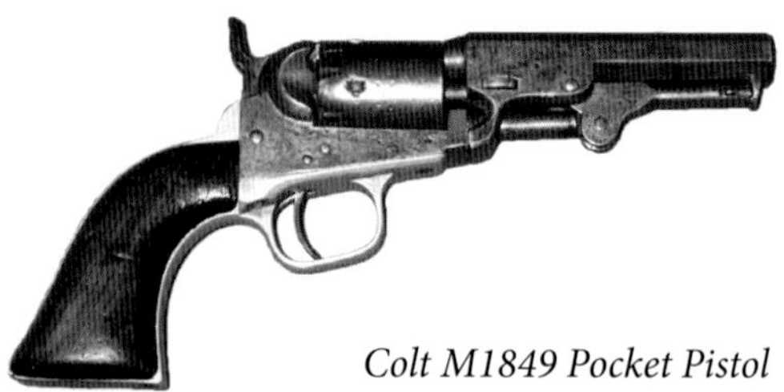

Colt M1849 Pocket Pistol

The inscription reads "J.B. Van Hagen"

Serial No. 176985 with the inscription "Nevada City, Cala".

and this 1849.31 caliber Pocket Model #176985 was there. The engraving on the back strap is contemporary to that period and indicates ownership by J.B. Van Hagen of Nevada City, California. Unlike some of the towns that sprang up during the Gold Rush and eventually disappeared, Nevada City still flourishes and maintains an authentic flavor of that era.

Van Hagen was a law enforcement officer from 1859-61, serving as sheriff of Nevada County. At the same time, he was the collector of foreign miners' license fees. As captain of the Nevada Rifles, a local volunteer militia organization, he served in a campaign against Indians in Nevada who were slaughtering white settlements. On June 20, 1860, he is recorded as having fought a duel with another member of the volunteer company, R.B. Moyes. Duels or gunfights in California at that time were fairly common.

In May of 2001, I visited Nevada City, California, specifically to do more research on this pistol and Sheriff J.B. Van Hagen. I was fortunate to meet Mr. Ed Tyson of Searls Historical Library Division of the Nevada County Historical Society. Mr. Tyson was about eighty years old and knew almost everyone in the county including family histories and stories. Another rich source of information is David Allen Comstock, author of numerous books on the area from the Gold Rush era from 1845 to 1869. In particular, his book, *Greenbacks and Copperheads 1859-1869,* contains numerous historical references to Sheriff John Bence Van Hagen.

Mr. Ed Tyson's research provided much of the following information: Van Hagen was a charter member of Oustomah Lodge No. 16 I.O.O.F., which was organized in Nevada City November 4, 1853. He was also a member of the Nevada Rifles, which was organized February 13, 1858 and became captain of the organization in 1860.

While a deputy sheriff, Van Hagen participated in a raid on a cabin in French Ravine, two miles from Grass Valley, California, on March 24, 1859. The posse of six men were intent on capturing three robbers that were supposed to be in the cabin. However, when they approached the cabin, eight men rushed out and opened fire on them, and the outnumbered posse retreated. Of the robbers, Eugene Whitney was killed, William Riley (AKA

Buckskin Bill) was wounded and the balance made good their escape. Later that year, Van Hagen was elected sheriff of Nevada County and served two one-year terms.

J.B. Van Hagen (artist's depiction)

His tenure as sheriff seems to have been rather eventful. He was involved in two very long, complicated lawsuits, both of which he lost. In the first one, he was sent as sheriff to the town of Orleans Flat to attach the store and wares of William Rauch. Van Hagen occupied the store for thirty-four days, during which time he seized a team of mules and other equipment. Mr. Rauch sued for loss of business and won a judgment against Van Hagen for $2,000, plus court costs.

Van Hagen appealed to the California State Supreme Court, where the appeal was denied and the case dismissed. Then the sheriff sued the two men in Sacramento who had put up the bond in case of a judgment against him.

In the meantime, the judge of the 14th District Court had ordered Van Hagen to apply $900, which he had deposited in his name in Sacramento, toward payment of the judgment against him. Upon his failure to comply, the judge ordered his arrest and required that he be brought to court.

The long case ended when the sheriff finally paid the $900, and the sureties in Sacramento paid off the rest of the judgment. Van Hagen did not go to jail.

The second lawsuit had to do with the disputed sale of a mining claim at Orleans Flat. At a sheriff's sale of the claim, the Eureka Lake Mining Company was the highest bidder at $1,330. However, it was the law that the property could be redeemed within six months of the sale. According to Sheriff Van Hagen, on September 19, 1861, Orlando Evans claimed the right to redeem the property and deposited the money with the sheriff. Eureka Mining Company disputed the legality of this transaction, and sued to force the sheriff to issue a deed to the property.

Judge Searls ordered that a Writ of Mandamus be issued Sheriff J.B. Van Hagen, commanding him to immediately issue the deed or appear in court on October 16, 1861 at 10 o'clock to show cause why he had not done so.

Van Hagen again appealed to the State Supreme Court, who again dismissed the case with the stipulation that the defendant, Van Hagen, pay all costs.

The Nevada Rifles Company was organized in Nevada City, February 13, 1858, with Henry Meredith, Captain; Phil Moore, First Lieutenant; George Story, Second Lieutenant; and Jerome Moore; Third Lieutenant. In the fall of 1858, Rufus Shoemaker became Captain, and in 1860 was succeeded by John Bence Van Hagen. The company served with honor; its members in 1858 were among the leading business and professional men in the city.

In regard to Sheriff Van Hagen's activities in the Nevada City Rifles, the Indian War in the Washoe Country in 1860 is of special interest to Nevada County, because of the prominent part taken in it by her citizens. On the evening of May 7, 1860, intelligence of the massacre of seven white men by Indians was brought to Virginia City. Two companies, one commanded by Major Ormsby and the other by Captain McDonald, in all over 100 men, proceeded toward the scene of the massacre, below the great bend of the Truckee River. They followed the trail until on the twelfth, near Pyramid Lake, Nevada, they were ambushed by a band of Paiutes in a pass.

The men fought desperately until their ammunition became exhausted, and then sought to escape by flight. Many were killed in the action, while many more were shot in their attempt to escape. Henry Meredith, an old and respected citizen and businessman of Nevada City, was with the party, and fell while fighting bravely after many had fled. The news reached Nevada City on Sunday; the alarm bells were rung, and the people assembled in the theater and made arrangements to send aid to the terrified settlers.

All that night men were busy making cartridges and preparing ammunition. Early in the morning a volunteer company of thirty men, under Captain J.B. Van Hagen of the Nevada City Rifles, started for the scene of action, having a great amount of ammunition and about sixty muskets. At Virginia City the company was increased by seventy-seven men, and served through the campaign of six weeks, doing good service. When the company returned from the seat of war they brought back the body of Henry Meredith and many others, which were received by the citizens of Nevada City in procession a few miles from the city, and buried in the city cemetery with honors and respect.

During the celebrated campaign against the Indians in Nevada in 1860, which included not only the Nevada City Rifles, but also a large number of volunteers from this county, a disagreement occurred between Captain J.B. Van Hagen and one his men—R.B. Moyes—in regard to some point of duty. Moyes decided that "when this cruel war is over..." he would demand satisfaction, and when the company returned covered with glory and dirt he issued a challenge to the offending Captain Van Hagen. This was promptly accepted by that valiant warrior, who decided that minnie rifles at sixty paces gave him about as good a chance for continuing in this world as anything else.

The fatal day was the twentieth of June 1860, and the gladiatorial arena was grizzly flat, in Yuba County, just across the middle Yuba River. The ground was paced off by the dignified seconds, sixty good and true paces. As the combatants took their stations with anger in their eyes and rifles in their hands, the sky glowed blood red with the just rising sun. Both fired at the word, and upon discovering that they were still sound in body if not in mind, both champions demanded another shot. A sarcastic individual suggested that they put telescopes on the rifles, but he was quickly squelched, and the work of death went on. Once more did tongues of flame leap from the angry rifles, and once more did the smoke lift from the field of carnage and reveal the virgin sod free from the contamination of blood. A "big talk" was then held, which resulted in an amicable understanding and both heroes were spared for future deeds of valor.

From the book, *Lives of Nevada County Pioneers*, the following timeline appears for John Bence Van Hagan:

John Bence Van Hagen (1824-) (Dem)

1824	Born in Ohio, the son of I.P and Sarah Van Hagen.
1846	Fought in the Mexican War.
July 4, 1849	Arrived in California.
July 1857	Chosen as delegate to the state Democratic convention, from Nevada City, California.
July 22, 1857	Was a Nevada County deputy sheriff; went to San Leandro to pick up a suspect.
Jan 26, 1858	Charter member of Nevada Rifle Company.
Feb 13, 1858	Elected orderly sergeant of Nevada Rifle Company.
March 6, 1858	Left Nevada City in pursuit of Eldred Northup, an escaped prisoner.
March 11, 1858	Returned with Northup.
April 10, 1858	Won first prize for shooting at the Nevada Rifles target practice.
February 22-23, 1859	Under Sheriff Van Hagen, picked up Alex Griffin in Sacramento and took him to San Quentin.
March 24, 1859	Took part in raid at Grass Valley.
September 7, 1859	Elected Nevada County Sheriff (DEM)
February 25, 1860	Chosen delegate to state Dem (Adm) convention at Sacramento.
March 14, 1860	Elected Captain of Nevada Rifles.
May 14, 1860	Led a company of 20 men from Nevada Rifles to Carson Valley after

	Paiute Indians ambushed a party of Indian hunters that included Henry Meredith.
June 20, 1860	Dueled with R.B. Moyes at Grizzly flat, Yuba County.
March 8, 1861	Hurried to Carson City after hearing his brother Charles had been blinded in one eye and was very ill as a result.
February 1862	Arrested by San Francisco deputy sheriff for contempt of court for not paying over certain money in his possession. Hit attorney John Anderson for having him arrested and brought to Nevada City.
August 30, 1862	Defeated as candidate for sheriff of Storey Co. Nevada (ND).
September 9, 1862	Audit shows $1,802.56 shortage in taxes collected by Nevada County sheriff in 1859-1861. Some thought stolen by deputy tax collector Phil Moore before he left town.
1875	Elected to Nevada state assembly from Lincoln County.
January 12, 1877	Member of the Mexican Veteran Association of the State of Nevada.

Apparently, Van Hagen left California soon after his second term of sheriff was completed. The local newspaper, the *Nevada Transcript,* carried an item noting that on June 24, 1869, J. Bence Van Hagen was in White Pine County, Nevada, and running for county recorder. Also, the *History of Nevada*, published in 1881 when Van Hagen was fifty-seven years old, carries three references to him. He is mentioned in connection with the Battle of Pyramid Lake; he is a member of the Mexican Veterans Association of the State of Nevada; and he is a member of the State Assembly.

Using the above information, it is reasonable to assume that J.B. Van Hagen probably died in the State of Nevada, but proof of this will have to await future research that may turn up an obituary. In any case, Sheriff J.B. Van Hagen had an eventful and interesting life in the early years of the Far West.

Chapter Sixteen

Engraved And Inscribed Smith & Wesson Revolver Leads to the Discovery of a Sinister Plot

Left side of the Smith & Wesson revolver Model 1-1/2 Top Break, .32 cal.

Inscription on the backstrap reads "John Baird".

In 1882, it appears some family members, lawyers, and doctors conspired to rob and perhaps kill John Baird, a wealthy retired vice-president of the Metropolitan Elevated Railroad Company of New York City. This plot was thoroughly investigated by the New York Supreme Court from 1880 through January 1887, but would not have resurfaced in 2007 unless John Baird's revolver came up for sale at an Ohio Gun Collectors Show on November 23, 2002.

I purchased this interesting revolver from Floyd R. (Bob) Everhart, one of the finest quality gun dealers in the country. At the time, Bob possessed no additional information on this gun, but I purchased it for its quality and beauty, hoping I could uncover more information about "John Baird."

Mr. Roy Jinks, Smith & Wesson's official historian and president of the Smith & Wesson Collectors' Association, researched this gun and confirmed it was shipped June 24, 1881, to M.W. Robinson, New York, New York, Smith & Wesson's largest distributor.

John Baird

Next, I contacted the New York Historical Society for information on John Baird. In 2003, I received a thorough response from Mr. Eric Robinson as follows:

> "There were about a half dozen John Baird's living in New York City during the latter half of the 18th Century, but the only one of means was a Scottish immigrant engineer who became one of the elite of New York society. When he was about fifty (50) years old he was in the New York Social Register and lived in a fine mansion at 324 Lexington Avenue. He had become a member of the posh Union League Club

Metropolitan Co. 9th Ave. elevated railroad, 1876.

in 1868. During 1886 and 1887 Baird appeared in New York City newspapers many times. The articles detail a trial to determine Baird's sanity, which was in question since 1882."

Mr. Robinson sent me twenty pages of articles from the *New York Times* and *New York Tribune* relative to Baird's life story, but at this point, I had no proof that this gun belonged to this "wealthy" John Baird instead of the other five "poorer" John Baird's. My reasoning was that a man would have to be fairly well off financially to purchase a gun described in the following letter from Mr. Everhart:

> "This letter is to confirm our transaction of 11/23/02 at the Ohio Gun Collector's Meeting as follows:
>
> Mr. Pitstick purchased one antique S&W Model 1-1/2 Top Break, .32 cal. Revolver, Serial #53582. This arm is richly deluxe Nimschke engraved originally, and in a most profuse full coverage manner. Left of frame is an oval portrait of a hunting dog, likely a favorite of the original owner, a Mr. John Baird, whose name is finely script engraved on the backstrap. This gun is equipped with a fine original set of pearl grips with nice coloring, generally excellent condition but with a flaw in the barrel. Finish is original nickel and about 40%-50% retained."

This gun probably cost about $100-$150 in 1882, when railroad laborers were making about $30 per month.

As a result, I decided to keep digging through information on this wealthy John Baird in the belief that he was very possibly the original owner of this revolver. It was in 2006, that I finally uncovered the whole story concerning the plot against John Baird, including the fact that in 1882 while on a spending spree he bought an engraved Smith & Wesson pocket revolver, probably because he feared for his life. This information was printed in the July 18th, 1886, *New York Times* as summarized below:

> "On petition from his brother, Hugh Baird, a Writ of Habeas Corpus was issued by Judge Barnard, to show cause

why John Baird should not be discharged from Stanford Insane Asylum. Hugh Baird is of the opinion that the patient is cured so that his normal mental powers are restored and that he is able to take care of himself and his fortune of about $1,000,000 (in 2007 dollars this is equivalent to $18,816,000). Hugh Baird was represented in court by the law firm of McClellan and McClellan and the patient's wife and other family members by Al Sullivan of Sullivan and Cromwell, who also represented Dr. Barstow, Superintendent of the asylum named. Before he was committed, Mr. Baird had been spending his money extravagantly, buying horses and carriages, silverware, drinking heavily, etc. He believes that at Delmonico's his wife tried to poison him. He once squandered $15,000 in two weeks.

"Dr. Barstow said the patient was suffering from chronic mania, with delusions. A commission had been appointed to inquire into his mental condition. The family desires a full investigation without delay.

"Attorney McClellan, in addressing the court on behalf of Hugh Baird, said he represented two brothers of the patient, Thomas and Hugh; both of whom assured him that John Baird is not only in good bodily health, but is also certainly able to take care of himself and his affairs. He has great ability, success in business, and counsel thought he ought to be discharged.

"Hugh Baird addressed the court, saying: 'I visited this asylum and said to Dr. Ferris that my brother ought to be out,' and Dr. Ferris replied, 'I thought so many months ago.' Hugh Baird privately expressed concern to Thomas that John had purchased an engraved Smith & Wesson revolver."

Following is John Baird's life story and self defense on the witness stand in his own behalf from the September 11, 1886, *New York Times*:

"Mr. Baird told his story to the jurors who are to determine his competence to take care of himself and his estate. Before Mr. Baird took the witness stand, his attorney, McClellan, said some of the acts that were the cause of the charge of insanity in 1882 were the results of stimulants administered to him by his doctors. As soon as the stimulants were stopped, he recovered his normal mental tone. Throughout the two hours of testimony, he never made a mistake respecting a date or a detail. He was born at a little place near Glasgow, Scotland in 1820 and learned the trade of a mechanical engineer in Glasgow. He married in 1842 and came from London, England to this city in 1844. After working here and elsewhere for eleven years, he became connected with the Cromwell Steamship Line. He designed and constructed the first propeller that went on the open sea. Prior to his venture, propellers were only used on Long Island Sound. When the elevated railroad was proposed he took an interest in it. To obtain $10,000 for investment he mortgaged his home on Lexington Avenue. He became General Manager, Vice-President and Construction Engineer of the Metropolitan Line.

"In June, 1879, his wife died, Mr. Baird said, and she was soon followed to the grave by their eldest son, James. He fell ill from these tragedies and was nursed by a friend of his wife. She cared for him in such a way that he believed she saved his life. On his recovery in 1880 he married her. She told him she had long been a widow. For two years their married life was perfect. In the spring of 1882 they were

Elevated steam train with cars.

Elevated railroad at 110th Street, ca. 1876.

starting a trip to Europe when Mrs. Baird told him that she was not a widow when she married him. This news perplexed him very much, but he took her to Europe with him. Before leaving he made a Will by which he bequeathed her the use of his house and one third of his income for her life. In Glasgow he was sick and on his return very feeble. He had joined the Presbyterian Church when he was young, and had always since been an earnest Christian. He denied that he had ever quoted to his children a verse from the Seventh Chapter of Micah devoting his enemies to hell fire. He denied also that he had insanely claimed to be of royal blood. He had merely pointed out that he was a remote relative of the Earl of Elpinstone. As to the delusion that he had been poisoned, Mr. Baird said he believed that because some food his wife had given him made him very ill. Once he heard his servant, Ben, tell two men who came to his room that it made no difference whether he took his medicine or not. He was doomed anyway.

"Mr. Baird said he made a new Will in the winter of 1885 which bequeathed his property to his six children to be divided equally, with the exception of $3,000 to be given to his sister. His second wife, Elizabeth T. Macrink Baird, was left entirely out of this Will.

> "Attorney, Al Sullivan cross-examined Mr. Baird. The latter said he believed his wife and son, John S. Baird, and Dr. Miner and Dr. Barstow along with their attorneys had conspired against him with the purpose of taking control of his entire estate. They would accomplish this by keeping him locked up in an insane asylum or even possibly killing him. Mr. Baird said he has seen no reason to change his mind about this conspiracy. This concluded his testimony."

Mr. Baird was shocked by his second wife's deception. This was a serious matter for two reasons: 1) John Baird thought she may be a bigamist and still married to William Macrink, and, 2) Even if Elizabeth Macrink Baird was divorced, John Baird's strict Presbyterian faith held Mr. Baird to be committing a great sin by cohabitating with a divorcee. This dilemma is held to be what made John Baird mentally sick in Glasgow and upon his return.

It is documented that John Baird had an excellent mind and had amassed a fortune of about $1,000,000 through his investment in the Metropolitan Company's Elevated New York Railway and other wise business decisions. In 1879, Vice-President Baird told the *New York Times* the total number of passengers carried by the Metropolitan Line daily was between 50,000 and 60,000 and the number of trains run each day was 441, a large and complicated business.

On December 28, 1882, Elizabeth Macrink and all of Mr. Baird's grown children had him committed through Judge Russell and Dr. Miner to Stanford Hall, Flushing Long Island, a private insane asylum, where he remained against his will for three and a half years. Meanwhile, John S. Baird, one of his sons, took control of his entire fortune.

An inquiry into his sanity was placed in the hands of a sheriff's jury. This commission was appointed by the New York Supreme Court. The jury met in session and found John Baird insane by a vote of fifteen of the eighteen jurors. However, Chief Justice Donahue reviewed the proceedings and concluded that a full and careful examination of the case "fails to convince me that the proofs show the party incompetent to manage his affairs and, therefore, the verdict is overturned." He was declared sane and set free.

This Declaration of Sanity confirms that John Baird had good reason to fear for his freedom and life itself. It appears that his second wife and grown children were, indeed, attempting to gain control of his money. Further proof of this is that his two brothers, Hugh and Thomas Baird, were the ones to recognize his sanity and get him out of this asylum. In addition, his brothers had nothing to gain but a just decision, since they were not his heirs!

The *New York Times* reported on January 29, 1887, that John Baird sued for an annulment of his marriage to his second wife. The grounds were that at the time of their marriage, Elizabeth Macrink was still married to William Macrink, and, therefore, she "is a bigamist and not legally able to marry." John Baird died peacefully of natural causes on October 17, 1891, at seventy-one (71) years of age.

Whenever I hold this beautiful revolver, I think of the family tragedy that occurred and that John Baird felt he had to purchase this gun to protect his life.

John Baird's gun in an original wood case.

Chapter Seventeen

The Guns of Caldwell, Kansas – A Notorious Chisholm Trail Cattle Town

Judge James Kelly's Remington Derringer inscribed, "Judge James Kelly, Caldwell, Kansas"

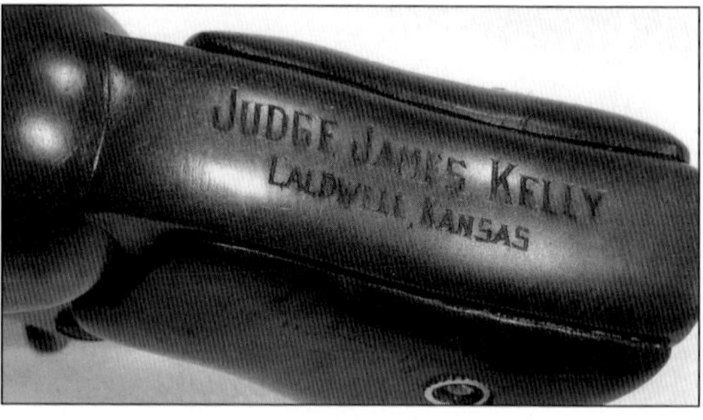

In 1996, I purchased a Remington Ring Trigger Four-Barrel Derringer from noted historian and author Charles G. Worman. On the backstrap is an inscription "Judge James Kelly, Caldwell, Kansas." Being very interested in history as well as guns, I had the opportunity to visit Caldwell in April 2002 after attending the Tulsa Gun Show. Prior to going, I had called the Carnegie Library in Caldwell and spoke with Ms. Norma White, head librarian, about my planned visit and research project related to Judge James Kelly and his gun. I told Norma I was planning to write an article for a national gun magazine on my inscribed Remington. Norma was enthusiastic and knowledgeable about Caldwell history, and offered to do some research prior to my coming. She also explained that a large portion of Caldwell citizens were local historians as well, and that

she would introduce me to some of them upon my arrival.

I have to tell you, that Norma White and the citizens of Caldwell greeted my wife, Sharon, and I like we were President and Mrs. George Bush! We couldn't believe the warm welcome and endless help we received from so many delightful people including Mr. & Mrs. Don White, who provided access and copies of their historic photos; Mr. Rod Cook, who researched the Internet and his personal files for information I needed; Norma White, who provided library access to historic newspapers plus a tour of both of Caldwell's museums; Mr. Colin Wood, author of, *Marshal's of Caldwell,* whose book provided much information; and Mr. & Mrs. Carson Ward, who owns part of the former King Ranch, told many enlightening stories and so on until I realized I had encountered a large story about a great and still authentic cattle town, and at least four historic guns that needed exposure to all of America.

Caldwell, Kansas, known as "The Border Queen," should be the most visited of the old cowtowns, (Dodge City, Abiline, Wichita, Ellsworth, etc.), because Caldwell still retains the size,

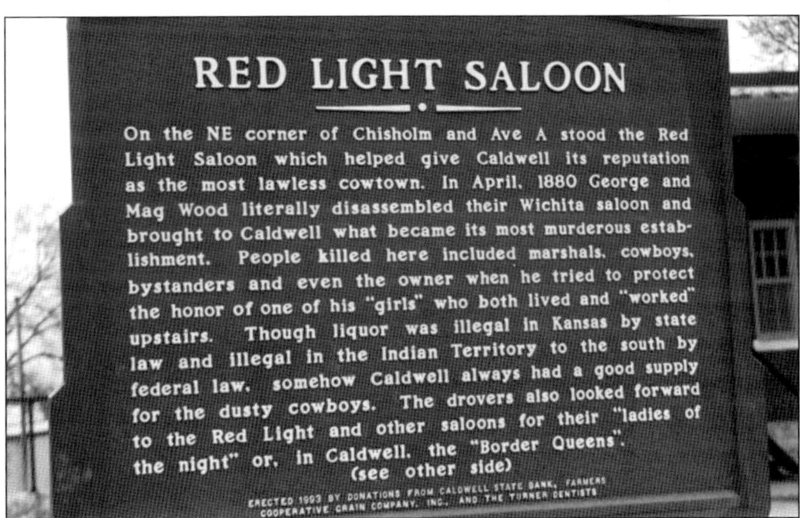

Leland Hotel, built in 1880.

flavor, and historic structures much as it was in the 1880s. Allow me to tick off just a few incredible facts relating to the significance of Caldwell's Old West history:

- In 1880, the Atchinson, Topeka and Santa Fe R.R. had its western terminus at Caldwell and by 1885 over four million Texas cattle had been shipped east from 1879 to 1885.
- Caldwell boasted twenty saloons, fifteen madams, and 119 prostitutes.
- In this six-year period there were 480 recorded arrests for gambling, 168 recorded arrests for prostitution and hundreds more for gun play, fighting, theft, drunkenness, and murder.
- In six years Caldwell went through fourteen Marshals — at least four were shot and killed in the line of duty.

Henry Brown, Marshal of Caldwell, Kansas

On the walking tour of the "Cowtown District", the following markers each tell at least one cowtown story concerning an event, person, or place. The project, started in 1993, has been funded completely with private donations in excess of $40,000. The markers were researched and written by local historians. New markers are planned and erected as donations are received. All markers are on display about five blocks of each other within the "Historic Cowtown District" area.

Historical Marker Locations

Address	Title
101 N. Main	Southwestern Hotel
17 N. Main	Marshal Henry Newton Brown
1 N. Main	Shooting Up Main Street
3 S. Main	Last Chance Saloon
23 S. Main	Leland Hotel
103 S. Main	Stock Exchange Bank
127 S. Main	Grand Opera House
124 S. Main	Legend of Mount Lookout
102 S. Main	Caldwell's History
30 S. Main	Native Stone Building
8 S. Main	Talbott Gang Shootout

Opera House, opened in 1884, burned in 1918.

23 E. Central	Prohibition Movement
203 E. Central	Railroads
12 E. Central	Last Land Rush
12 N. Main	Marshal George Flatt
16 N. Main	Red Light Saloon

Other great points of interest are located nearby:

- "Ghost Riders of the Chisholm Trail" — A life-sized silhouette of a cattle drive astride the actual Chisholm Trail. A historic marker at the site relates the story two miles south of Caldwell on U.S. 81.
- "Those Who Came Before" — A limestone relief by local Cherokee sculptor, Eddie Morrison, depicting the cultures who have touched this area, one-half block east of Main Street on Central.
- "Cowboys Driving Cattle" — A mural by artist Kenneth Evett, a student of Thomas Hart Benton. This is on the south wall of the Caldwell Post Office, 14 N. Main Street.
- "Heritage Park Mural" — A charming depiction of the historical progression of life on the plains by local artist, Brenda Lebeda Almond, located on the south side of Heritage Park at 102 South Main Street.

Caldwell, Kansas, was incorporated as a City of the Third Class in July 1879 and elected the following officers on August 7: N.J. Dixon, Mayor; James D. Kelly, Sr., Police Judge; J.A. Blair, F.G. Husson, H.C. Challes, A.C. Jones, and A. Rhoades, Councilmen. Mayor Dixon appointed George Flatt as Caldwell's first Marshal on August 21, 1879, and James D. Kelly, Jr. was appointed City Clerk. J.D. Kelly, Jr. was also the first editor and founder of the *Caldwell Post Newspaper.*

The Remington Ring Trigger Four Barreled Derringer was carried by Judge James D. Kelly Sr. for many years (see backstrap inscription), because Caldwell was such a dangerous town and the nature of being a police judge over such violent hell-raisers was risky business. James Kelly Sr., his wife, Rebecca, and seven children arrived in Caldwell, Sumner County, Kansas, in 1876 from

Kentucky. They had come by way of Chicago on the Fort Wayne and Pennsylvania Railroad looking for a better life and new opportunities. Mr. Kelly was a lawyer and justice of the peace and set up offices in partnership with lawyer, T.H.B. Ross, on the west side of Main Street late in 1876. He conducted a general law practice and became involved in city and county politics. Elected Police Judge in 1879, he soon became a very busy man as revealed in a study by Rod Cook of the "Border Queens" police dockets from 1879-1885.

Most of Caldwell's rowdiest history is to be found within this period. At that time, Caldwell was the archetypal cowtown. Rough and crusty, it existed for cattle and cowboys. It was the cowboy's reward at the end of a painfully long hard trail. For them, it was the time and place to celebrate, and Caldwell had ample establishments that catered to celebrations.

Caldwell was located on the old Chisholm Trail and has been during all its history a "Cattle Town." Consequently, it has always had in and about it a class of people who have caused many scenes of disorder and bloodshed, which, if they were chargeable to the county, would not give it a good name by any means. From its location, it has always been a favorite resort of the cowboy and the desperado when off duty, and in its earlier days was a decidedly fast and dangerous locality.

The cowboys, together with a contingent of resident hell-raisers, kept the lawmen and judge busy throughout the era. These wild, often violent times ended in the last days of 1885. The town was settling down. The cattle trade was starting to slack off and the city was trying to curtail gambling, close the saloons, and become more temperate. January 1885 saw the last of the professional gamblers, and prostitution was eradicated by that year's end.

An examination of the Caldwell Police docket records for these turbulent times reveals a great deal. The pages were numbered 1 through 450, and each case was entered on a successive page. Confusingly, the cases were not entered in chronological order. That is, the date the case was entered into the book was most likely not the date of the offense. For example, a case dated July 28 on page 51 might be followed on page 52 by a case dated July 2. Apparently, selected trusted offenders were allowed to postpone their appearance before the judge until they had the money to pay the fine and costs. They were, therefore, granted the courtesy of avoiding jail time. Another reason for the mixed dates is that many times the arresting marshal collected the fine and costs himself and turned them into the judge at a later date. Such was the case of the last two arrests made by Marshal Henry Brown. He made the arrests, but was killed before he had a chance to turn in the money.

Except for a short illness when James M. Thomas filled in, Kelly was police judge until early April 1883. Thomas Hart Benton "Bent" Ross then held the position until Kelly assumed it again in January 1885 through 1887.

Sometimes "handles" were entered in lieu of the offender's correct name. For example: Texas Dan Jack, Dutchy, Mr. Love, Soda Pop Jack, High Ball Joe, Dutch Fred, Red Shirt, Slim Jim, Keno, Cousin Carey and Little Jonnie were some of the names entered as defendants. Sometimes the Judge, not knowing the offender's name, simply did the best he could, i.e. "One Shorty," "Negro Cole," "John Mack whose true name is unknown," "Stranger," "One Taylor," "Colored Sergeant," "Innkeep," and many times of "Name Unknown."

The normal routine was to hold the violator in jail until the fine and costs were paid. The judge wrote, "committed to city prison until fine and costs are paid" or words to that effect on most pages. To add to the confusion, occasionally the disposition would be shown as "sent to prison." At times that meant that the person was sent to Wellington, but most of the time it meant Caldwell Jail. Jailed prisoners were allowed to work off their fines maintaining city streets and other odd jobs around the town site.

Many of the offenses were mundane —as we would normally expect for a cowtown – drunk and disorderly, disturbing the peace, petty theft, etc. Some of them were very funny, some were violent, and some were just curiously interesting.

Following is a cross section of offenses:

"Did threaten to shoot J.J. Snow."

"Confined for swearing in the court."

"Did draw a knife."

"Ride horse on sidewalk."

"Obscene and scandalous language."

One Shorty, "Made threats with a razor."

Joe Weideman, "Did resist arrest."

"Received stolen property."

"Indecent exposure."

"Did draw a knife and revolver upon David Swan."

"Did utter profane language in loud and boisterous manner."

"Draw a pistol upon Frank Higgins."

"Led a horse upon the sidewalk."

"Did carry a knife."

"Did make a great noise upon the streets."

"Did carry and draw a revolver upon Joseph Spiken."

Frank Pratt, "Did ride his horse upon the sidewalk. Prisoner was committed to the new city prison but escaped through the bars."

Peter Egan was "Drunk and disorderly. The Marshal finding out that the defendant has no money and that it would be an expense on the city for his board he turned the prisoner loose."

Thomas J. Ingrahm was arrested by James Johnson for "Carrying a revolver and threatening to use it on James Johnson." "Ingrham (sic) actually snapped it three times in Johnson's face. Lucky for Johnson, it misfired all three times. Perhaps the charge should have been attempted murder."

Included in the books between 1879 and 1885 were fifty-six assaults, 240 drunk, 480 gambling, 168 prostitution, seventeen "did shoot off a six-shooter," 146 "disturb the peace and quietude" or "did curse, swear, quarrel and use violent threatening indecent language," fifty-five "did carry a six-shooter" or "did carry concealed weapon," and 113 were "fighting." These were the known offenders — how many there were that didn't get caught is anyone's guess.

After 1887, Judge Kelly retired from politics and resumed his private law practice. James Kelly died at about 64 years of age in 1889, and was buried in the Caldwell Cemetery.

Henry Newton Brown, Marshal of Caldwell, and Deputy Marshal Ben Wheeler

As Marshal of Caldwell, Henry Brown found his most satisfying role and lifestyle. He was well-liked and widely respected as a dangerous gunman and an efficient officer. He found surroundings in bustling Caldwell more pleasant and comfortable than those he had experienced in the ranch country of Colorado in sparsely-settled Lincoln County, in the Panhandle Plains of Texas, or in the unsettled Indian Territory. During his tenure as city marshal, the only two men who were shot and killed in Caldwell died at the hand of Brown himself, and the once wild railhead remained tame under his supervision. Henry even took a wife and became a property holder, and he should have been a likely candidate to serve as a legendary Westerner and lawman for two or three decades. City Marshal Brown was paid the rather handsome salary of $100 per month. Brown's deputy, Ben Wheeler, was paid $75 monthly, a fifty per cent increase over the 1881 Marshal's salary.

Brown and Wheeler both had quiet and restrained personal habits, both had desperate backgrounds, which they wanted to keep hidden, and both could handle a gun. Wheeler was Marshal Brown's first choice as a deputy. Moreover, it was observed during his initial tenure on the police force that, "Mr. Wheeler has the sand, so the boys say, to stay with the wild and wooly class as long as they are on the war path." When Henry Brown was given a permanent appointment as City Marshal, Ben Wheeler was named Assistant Marshal.

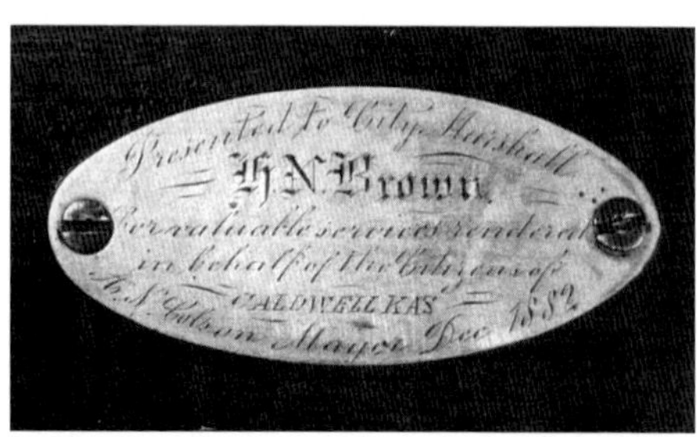

The inscribed plate on Brown's Winchester reads: "Presented to City Marshall / H.N. Brown / For valuable services rendered / in behalf of the Citizens of / CALDWELL KAS / A.N. Colson Mayor 1882.

A week later Brown received an expensive and lasting gift. Just as they had done with Bat Carr, local

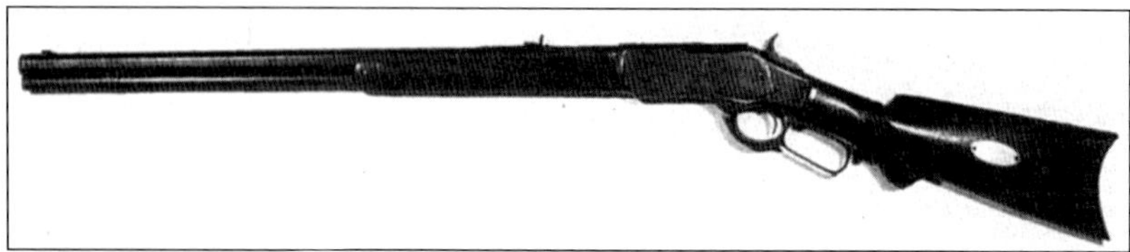

The citizens of Caldwell presented this engraved Winchester to Henry Brown on January 1, 1883. Courtesy of the Kansas Historical Society

Caldwell leaders raised a subscription to impress upon their new police chief how strongly they supported him and the order he represented. On Monday, New Year's Day, 1883, an unsuspecting Brown was lured into the York-Parker-Draper Mercantile store, where several friends already had gathered. Frank Jones stepped forward, made a speech appropriate to the occasion and the astonished Brown was presented with a magnificent .44-40 Winchester. The stock had been shaped from polished black walnut and sported a pistol grip, and the butt end was covered with intricately engraved gold plate. The barrel was octagonal, and on the right side of the stock was a plate ornately inscribed: "Presented to City Marshal H.N. Brown for valuable services rendered in behalf of the citizens of Caldwell, Kansas — A.N. Colson, Mayor, December 1882."

After Brown and Wheeler had been in charge of policing the town for a month, the Marshal decided to take a vacation. Following an absence of nearly a decade, he wanted to visit Phelps County, Missouri, and his relatives, show them his engraved Winchester, and impress them with his respectable position. In mid-winter the cattle buyers were gone and Caldwell was quiet, and Ben Wheeler was obviously capable of policing the town during such a lull. In fact, the *Caldwell Commercial* observed that, "Ben is equal to the occasion, being of that class of men who have very little to say but are very prompt when action is necessary."

Although Henry's opinion of himself as being superior to his Caldwell predecessors may have been tinged with exaggeration, when he returned to Caldwell he attempted to earn its justification. During his stay in Phelps County, he constantly wore his holstered six-guns and cowboy attire, despite the fact that he usually walked the streets of Caldwell clad in a suit and string tie. Apparently, he was trying to spread the impression that the Missouri farm boy who had left home eight years earlier had become a veteran plainsman: cowboy, lawman, gunslinger — a Westerner to be respected and envied, just as he had envied adventurous frontiersmen as a boy. Overall, his reserved manner had not changed much.

Henry's only sister, Ellen, was then residing at Cold Spring. Brown must have enjoyed showing off his Winchester and cowboy garb, and surely he opened up his past, to some extent, to his closest relatives. Perhaps he engaged in target practice with his six-gun, and maybe some of the boys were allowed to fire the cherished Winchester. Possibly, he confided to his sister or uncle or an older cousin about Billy The Kid and the Lincoln County killings and the fatal shootout in the Panhandle, and conceivably he may even have whispered about events that have never since been revealed.

Ben Wheeler, Henry Brown's deputy

In Caldwell, he received a welcomed homecoming. The week after Brown moved back into his room at the Southwestern, the *Caldwell Commercial* stated: "Since his return, the boys are not quite so numerous on the street at night." Apparently, Ben Wheeler had not quite lived up to expectations, but the constant patrol of the steely-eyed Brown

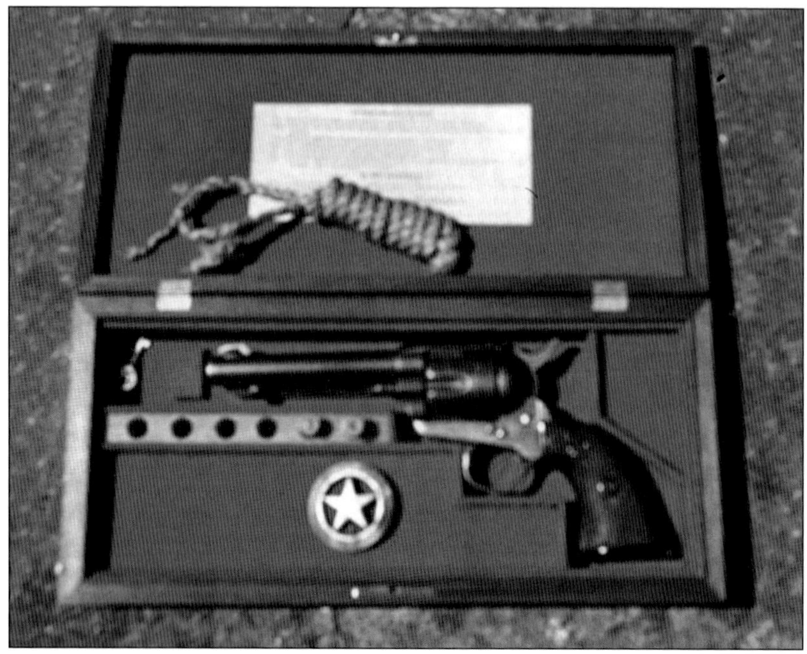

Deputy Marshal Ben Wheeler's gun

quickly settled the local toughs.

Things stayed so quiet for a time that spring promised to be very pleasant. The local schoolmaster, Professor Sweet, organized an outing to nearby Polecat Creek on the third Sunday in March. A somewhat shy participant in community socials, Marshal Brown went along as "guard" of the group. The picnic was spread beneath the shade of cottonwoods and elm trees that lined the creek banks, and amidst the laughter and horseplay, Brown must have experienced a feeling of acceptance and well being.

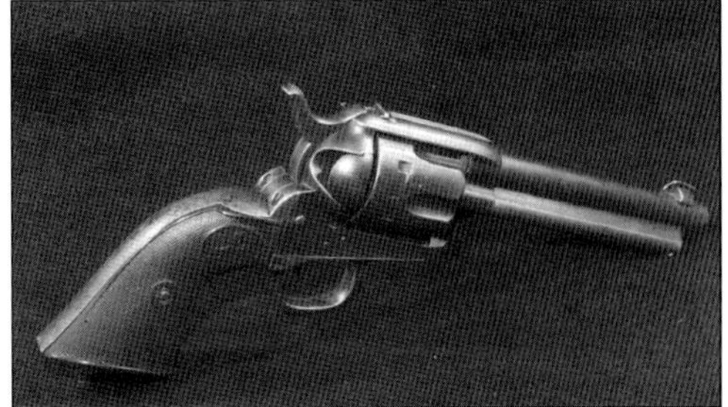

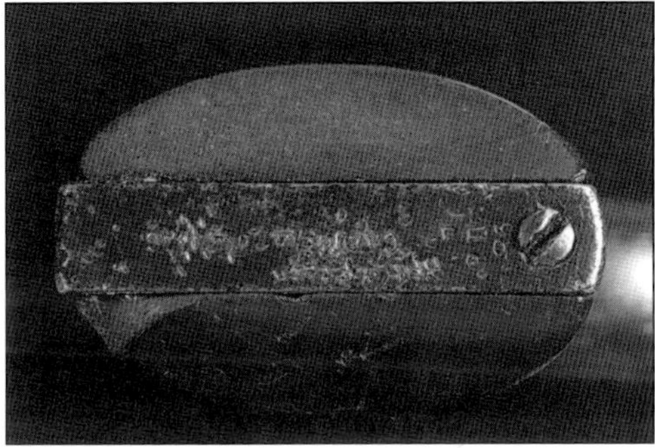

Butt of Ben Wheeler's gun with his name scratched in it

Early each April, the Caldwell city elections were held, and the newly elected City Council of 1883 wasted no time in reappointing Brown and Wheeler for the next term.

A few days later the local officers were called upon to assist in chasing horse thieves. Cash Hollister, a fearless Caldwellite who had been arrested several times for fighting — once while serving as Mayor in 1879 — somehow obtained an appointment as a Deputy U.S. Marshal. Almost immediately he arrested a horse thief, and not long afterward, was approached by another victim of rustlers. On Sunday, April 8th, J.H. Herron rode into Caldwell, sought out Hollister, and asked his help in capturing a gang of horse thieves he had trailed all the way from Texas. Herron had lost a pair of horses and a pair of mules, and the band had stolen animals from other individuals as well.

Hollister saddled up and rode with Herron toward Hunnewell, twenty miles east of Caldwell. The rustlers were camped a few miles southeast of Hunnewell, but a little scouting revealed that there were too many of them to safely apprehend. They were led by a rogue named Ross, and backed by Ross's two sons. Three females were present — Ross's wife, daughter, and daughter-in-law — and the latter woman had a child. Another group that had accompanied the Ross family was camped nearby. Hollister apparently left Herron to observe the rustlers, then rode back to Caldwell to enlist aid.

He arrived back in town on Tuesday, and by eleven that night Henry Brown and Ben Wheeler were headed toward Hunnewell with Hollister. In Hunnewell the lawmen added City Marshal Jackson and Sumner County Deputy Wes Thralls, and in the dead of night the posse continued on toward the sleeping camp.

By daybreak the officers had surrounded the Ross camp, and when everyone was in position, the quiet of dawn was abruptly shattered by shouted demands for surrender. The Ross party replied without hesitation, firing their Winchesters and drawing a hail of return fire. For half an hour there was a fierce exchange of shots, during which the older Ross boy was killed and his younger brother hit by two or three bullets. Gunfire from the camp stopped, and the posse moved in on the remaining rustlers. The wounded Ross brother proved talkative, indicating that the gang had headed north from Texas with about forty head of mules and horses, including a prized stallion for which a reward of $500 had been offered. He added that two members of the original gang had driven part of the stolen herd toward Wichita on the previous Sunday.

The dead rustler was left in Hunnewell, and the posse escorted the rest of the family to the county jail in Wellington. Anxious to return to their responsibilities, Brown and Wheeler apparently headed straight back from Hunnewell, arriving in Caldwell at eleven a.m. on Wednesday, the morning of the shootout. They immediately gave the editor of the *Commercial* details of the fight, just in time for inclusion in the Thursday edition of the weekly.

Then in the spring of 1884, the president and the cashier of the bank at Medicine Lodge were killed during a holdup. Citizens, quickly capturing the gang in Wheeler's Canyon, were astonished to discover that the bandit leaders were Brown and Wheeler. The prisoners were jailed but that night a mob came to lynch them. The sheriff resisted. In the confusion, the bandits ran. Brown was killed by several rifle bullets and a charge of buckshot. Wheeler was wounded and recaptured with two other accomplices, Billy Smith and John Wesley. He was hanged from an elm tree.

Terwilliger

Thus came the end of Marshal Henry Brown and Deputy Ben Wheeler!

Another great gun of Caldwell is displayed at Caldwell's Cherokee Strip Museum. The .41 caliber Colt Lightning and shoulder holster were carried by Frank Albert Terwilliger. It's the same type of gun Billy the Kid liked. The gun was primarily used to intimidate people, rather than for killing. Most cowboys were

Train going down from Caldwell through the Cherokee Strip at opening, September 16, 1893.

just good, hard-working people trying to stay out of trouble and make a living.

Born during the Civil War on June 7, 1864, in a house that is now a historic site, Frank Terwilliger left home at the age of 14 with $50 in his pocket. He made seven drives along the Chisholm Trail over a fifteen-year period, bringing up herds of cattle or horses from Mexico.

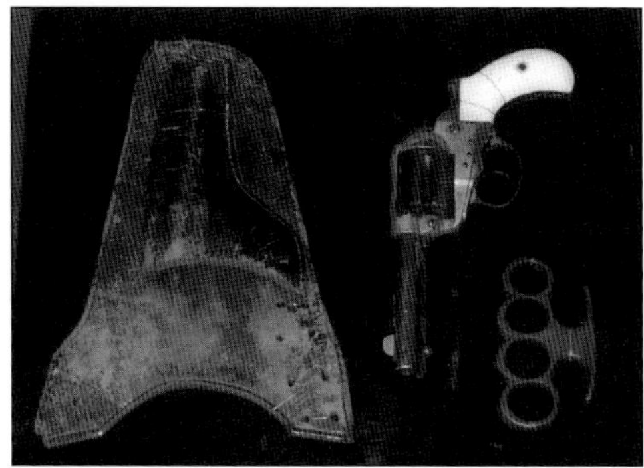

Terwilliger's gun: a Colt Lightning and Shoulder Holster

He was among those who raced for a land stake when the Cherokee Strip opened in 1893. He staked a claim to a piece of property, then bedded down for the night. The next morning he walked over the hill and came across a man who had staked a claim to the same piece of land on the other side of the hill. They flipped a coin to determine ownership, and Terwilliger lost. He returned to Kansas and bought some land with the $40,000 he had earned from the cattle and horses he had brought up the Chisholm Trail. But, the farm never flourished. On one of his last trips along the Trail, several cattle became mired in quicksand along the Red River. Terwilliger exhausted himself pulling the cattle to safety in the oppressive heat and he was never the same after that. His health slowly failed, and he died in 1903 at the age of 38. He left behind one son and a wife who was so devoted to him that she never remarried.

This completes the tale of the fourth documented and famous Caldwell guns. It is truly remarkable that a town of this size could produce so many great "Old West" stories. When you have a chance, be sure and spend some vacation time in historic Caldwell, Kansas.

Chapter Eighteen

Duels, Honor, and the Confederacy Major General John C. Breckinridge and His Horse Pistols

Several years ago I sold my only pair of dueling pistols, and soon realized I missed them. I called my publisher, Stuart Mowbray — a noted collector of dueling pistols — for advice on locating a good pair, which he freely gave and wished me good luck. His wish for me would come true in spades!

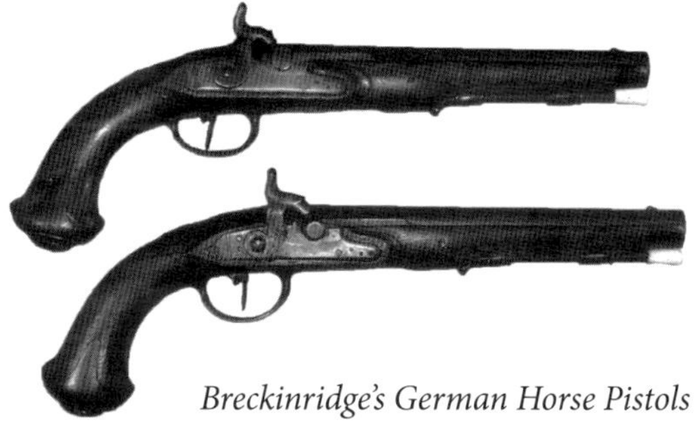

Breckinridge's German Horse Pistols

Barrel showing gold inlaid master's mark and inlaid maker's name

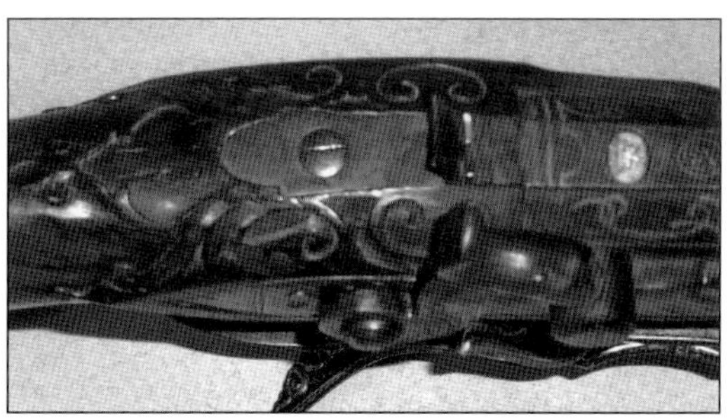

Handcarving at the breach

Next, I called my friend, Charles Layson, an excellent gun dealer in Lexington, Kentucky, and owner of Antique and Modern Firearms Company. It was a fortunate and surprising day for me because Charles knew I collect pistols where the owner is known and an interesting story may be unearthed through research. Charles's first words were, "I think I have a pair of dueling pistols just consigned to us that will intrigue you." The description at the time of my purchase was:

> "This German-made pair of 18th Century dueling pistols have, until recently, been in the possession of one of Kentucky's most prominent land grant pioneer families of Woodford County. They were originally the property of Edward (Ned) Blackburn, who's home, Equina, was located near Spring Station, Kentucky. They were used in a duel in 1855 by Ned's son, James W. Blackburn, who was seriously wounded in the leg, but recovered sufficiently to enlist in the Confederate Army in 1861 and serve four years until taken prisoner in 1864. He was listed as a farmer and attorney in the Census of 1870, and was elected to the State Senate in 1875. Before he was able to take office, he had to receive a pardon from the Kentucky Governor, since Kentucky does not allow a known duelist to hold state office. He served as Secretary of State from 1880 to 1883, and as such, signed a pardon for Miss Belle Brezing, Lexington's most famous Madam. The guns were made by Johann Andreas Kuchenreuter of Regensburg, Germany, around 1790 to 1795. The Kuchenreuter dynasty of gun makers was one of Europe's most famous for eight generations, from 1670 to 1898. While they crafted firearms of all types, and designed for many uses, their matched pairs of dueling pistols were considered to be the top of the art form. Both guns are in excellent working condition and were converted from flintlock to percussion around 1840. Seldom are such quality antique guns found with such strong provenance."

This was the information available at the time provided by the seller, Mrs. Elizabeth Fielder, Executrix and heir of the estate of her uncle, Joseph Blackburn Wise of Midway, Kentucky. An interesting sidenote about the above-mentioned Madam, Belle Brezing — she opened her third and best brothel at 59 Megowah

Street, Lexington, Kentucky. It was lavishly appointed and decorated in Victorian style. Belle's was not the only brothel on the hill, but certainly the most expensive and popular. Brezing attracted clientele from all over the nation who visited Lexington, known for its horse breeding and racing industry. She eventually appeared as the character, Belle Watling, in Margaret Mitchell's *Gone With The Wind,* which was confirmed by John Mitchell after Margaret's death.

Elizabeth Fielder presenting the pistols to Jerry Pistick and his wife.

This connection with one of Kentucky's most prominent families enamored me, so I purchased the guns immediately for the asking price. James Blackburn's brother, Luke Blackburn, was the Governor of Kentucky from 1880 to 1883, and four Blackburn relatives currently held positions in Kentucky's government, including state auditor. The price of $6,500 seemed fair based on Norm Flayderman's *Guide to American Firearms* regarding dueling pistols. He states if there were a documented duel and the duelists were known, "The guns are valued a great deal higher depending on the actual circumstances and people involved."

My research soon turned up revolutionary facts, far more important than the original information available at the time of purchase, August of 2011. Elizabeth Kennedy Fielder was the Executrix and genealogist of her illustrious family. I asked her to help me put together an affidavit of the history of the pistols and her family. Elizabeth was a terrific help! She wrote on September 17, 2011:

> "My grandmother was raised by her grandfather, U.S. Senator Joseph Clay Stiles Blackburn at the horse farm now

known as Fairview Farm, Woodford County, Kentucky. My grandmother, Terese Blackburn Lane Wise Rouse, raised her first son, Joseph 'Buddy' Blackburn Wise and my father at Fairview Farm. My grandmother kept journals most of her married life. One of those journals covers the year 1941.

"On August 24, 1941, my grandmother wrote in her journal the following: 'Buddy comes back about 9:30. He went to Danville to get the two old dueling pistols that used to belong to John C. Breckinridge. Uncle Jim Blackburn was crippled by one of them in a duel over a woman. Buddy wanted to own them because of the family association.' Now you can assert with some evidence that the pistols belonged to John C. Breckinridge who sold them to my uncle, Joseph Blackburn Wise, the grandson of U.S. Senator Joseph C. S. Blackburn, and the grand-nephew of James W. Blackburn who dueled against Dr. Theopilus Steele on February 16, 1855."

Dr. Steele was married to Edith Breckinridge, the sister of John C. Breckinridge. At the time of the duel, both James Blackburn and Dr. Theo Steele were only twenty years old and nei-

Sketch of the duel between James W. Blackburn and Dr. Theo Steele.

ther owned a set of dueling pistols. So, they borrowed John C. Breckinridge's pistols for the affair of honor. The duel was fought near Moreland's Tavern, about nine miles from Lexington, on the line of boundary of Bourbon and Fayette Counties. James Blackburn was wounded in the thigh, but Mr. Blackburn won the woman's heart and subsequently married Henrietta Everett. In addition to the woman, there was a long-standing family feud between the Steele's and the Blackburn's. In 1842, Captain Thomas Steele killed William E. Blackburn at Harmony Meeting House in Woodford County, Kentucky. Blackburn, on that occasion, attacked Steele, who was acquitted. This duel put an end to a long-cherished family animosity.

John C. Breckinridge, Vice President of the United States

Many readers may know who John C. Breckinridge was, but for those unfamiliar, he was a true Renaissance man. The highlights of his life are as follows: The American soldier and politician, John C. Breckinridge, was born near Lexington, Kentucky, on January 21, 1821. He was a member of a family prominent in the public life of Kentucky and the nation. His grandfather, John Breckinridge (1760-1806), who revised Thomas Jefferson's draft of the "Kentucky Resolutions" of 1798, was a United States Senator from Kentucky in 1801-05 and attorney general in President Jefferson's cabinet in 1805-06. His cousin, William Campbell Preston Breckinridge (1837-1904), was a Democratic representative in Congress from 1885 to 1893. Another cousin, Joseph Cabell Breckinridge (b. 1842), served on the Union side in the Civil War, was a major general of volunteers during the Spanish-American War (1899), became a major general in the regular United States Army in 1903, and was Inspector General of the United States Army from 1899 until his retirement from active service in 1904.

John Cabell Breckinridge graduated in 1838 at Centre College, Danville, Kentucky, continued his studies at Princeton, and then studied law at Transylvania University in Lexington. He practiced law in Frankfort, Kentucky, in 1840 and in Burlington, Iowa, from 1841-1843, and then returned to Kentucky and followed his profession at Lexington.

In 1847, he went to Mexico as Major of the 3rd Kentucky Volunteers. In 1849, he was elected a Democratic member of the Kentucky legislature, and from 1851-55 served in the national House of Representatives. President Franklin Pierce offered him the position of Minister to Spain, but he declined it. In 1856, he was chosen as Vice President of the United States on the James Buchanan ticket, and although a strong pro-slavery and states' rights man, he presided over the Senate with conspicuous fairness and impartiality during the trying years before the Civil War.

In 1860, he was nominated for the Presidency by the pro-slavery secessionists from the Democratic National Convention, and received a total of seventy-two electoral votes. As vice president and presiding officer of the Senate, it was his duty to make the official announcement of the election of his opponent, Abraham Lincoln. He succeeded John J. Crittenden as United States Senator from Kentucky in March 1861, but having subsequently entered the Confederate service, he was expelled from the Senate in December 1861.

As brigadier general, he commanded the Confederate reserves at Shiloh, and in August 1862 was promoted to major general. On the 5th of this month he was repulsed in his attack on Baton Rouge, but he won distinction at Stones River (December 31, 1862 to January 2, 1863), where his division lost nearly a third of its number. He took part in the unsuccessful siege of Vicksburg under Major General Van Dorn in July 1863. At the Battle of Chickamauga, he defeated General Franz Siegel at Newmarket, Virginia, on May 15, 1864, and then joined Robert E. Lee and took part in the battles of Cold Harbor on the 1st and on the 3rd of June. In the autumn he operated in the Shenandoah Valley, and with Early was defeated by Philip Henry Sheridan at Winchester on the 19th of September.

Being transferred to the department of Southwest Virginia, he fought a number of minor engagements in eastern Tennessee,

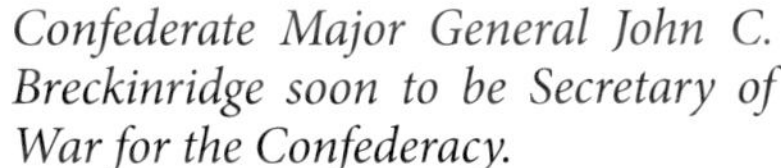

Confederate Major General John C. Breckinridge soon to be Secretary of War for the Confederacy.

John C. Breckinridge, United States Senator

and in January 1865 became Secretary of War for the Confederate States. At the close of the war he escaped to Cuba, and from there went to Europe. In February 1869, having received a general amnesty from President Andrew Johnson, he returned to the United States and resumed the practice of law at Lexington, Kentucky, where he died on the 17th of May 1875. In honor of Breckinridge, Lexington erected a full-sized statue in their main downtown park. Very few people led such an illustrious and tumultuous life.

The above enumerates the facts of Breckinridge's life, but doesn't unveil the real man behind the action. A number of quotes from his peers will richly illustrate the true essence of his character and integrity.

His friend, Thomas Crittenden, declared, "That for brilliancy of parts, readiness of wit, a happy force and luxuriance of imagination, acuteness in argument and felicity of expression, as a gentleman, scholar, statesman or a companion, no one could be compared with him. In addition, his modesty is exceeded only by his worth." Other friends remarked on his "tall and manly person, graceful carriage, polite address, impressive countenance and dignified and imposing presence."

During the Mexican War of 1847, his commanding officer wrote a correspondent, "Major Breckinridge is the soul of chivalry — bold, brave, and dignified — and still very affable in his manners. He is a man, too, of great energy and perseverance — firm in his commands, but lenient in his requirements."

During the Mexican War, Major Breckinridge was known to loan his horse to wounded men while he walked alongside tending to their needs. Thanks to his constant concern, his popularity with his men soared. He was easily the best-liked man in the regiment. One soldier wrote home, "Of courage, talent and all the generous qualities requisite to make a noble soldier, Breckinridge is in an eminent degree possessed."

During the Civil War, Major-General Reuben Davis enjoyed Breckinridge's company and wrote, "He was a goodly sight, sitting on a stool or table, with a glass of Old Shuck in his hand, and that grand voice of his vibrating through the tent like a deep-toned bell. He was not only a most elegant gentleman, but also genial and full of spirit, and ready to meet the worst of days with a sort of gay courage that sat well upon his stalwart manhood. As he sat in his saddle, he seemed altogether the most impressive-looking man I had ever seen."

Of all the Confederate Secretaries of War, John C. Breckinridge was the one most suited, from all vantage points, for the office. He came with the stature to give dignity to his office, command the respect and affection of the people, and stand up to President Davis without fear. At the same time, he had the courage and foresight not to be a conventional administrator. By the time he took over, the war was lost, but the way he thereafter worked toward the single goal of an honorable demise, shows his great ability to face a tough situation squarely and devise a plausible plan of action. We can thank Breckinridge for preventing the destruction of the Confederate archives.

It is my belief the original owner of these pistols was John C. Breckinridge's father, Joseph Cabell Breckinridge, who was born in 1788 and died in 1823 at the youthful age of thirty-five. Joseph served in the War of 1812 and may well have purchased these horse pistols specifically for his use in the war. Joseph was in at least two campaigns along the future route of the Illinois and Wabash Railroad Lines in 1813.

While there is no direct proof, there is a definite possibility General John C. Breckinridge used this pair of horse pistols during the Civil War, given the shortage of weapons available to the Confederacy. On page 535 of the book, *Breckinridge, Statesman, Soldier, Symbol,* by William C. Davis, is an account of his escape from Florida to Cuba in 1865. "Mr. Wood, Russell and O'Toole would sail out to meet the Federals. Breckinridge hid in the bushes, pistols in his hands, while they quieted the enemy's suspicions." *Could these pistols be the ones featured here?* Probably not, but it is possible!

John C. Breckinridge died in 1875 and one of his descendents, Eugenia Young of Danville, Kentucky, was his heir. Eugenia remained unmarried all of her life and died in 1940 at Danville. Her estate was settled in 1941, which ties in with the purchase of the pistols on August 21, 1941, by Joseph Blackburn Wise. Miss Young's estate and the Rodes' family also provided funding for the Boyle County Public Library in Danville. Mrs. Julie Rodes of Danville, Kentucky, a living Breckinridge descendent, is currently in possession of the Eugenia Young Estate papers, which she verified in January of 2012.

The discovery of these amazing pistols was more than just "serendipity" and great fortune in gun collecting. The ownership of these guns gave me the opportunity to hold history in my hands and relive, through Confederate eyes, an important part of our nation's past. *What other hobby can provide such an experience?*

References

1. Davis, William C. (1974), *Breckinridge, Statesman, Soldier, Symbol.*
2. *Antique and Modern Firearms Company Information*, by Charles Layson
3. Affidavit dated October 21, 2011 by Executrix, Elizabeth Fielder.
4. Jamie Helle's research from The Boyle County Public Library, Danville, Kentucky.
5. Eicher, John H., *Civil War High Commands.*
6. The National Archives and Library of Congress, research by Vonnie Zullo.

Note This article is based on the information available but, as with all history, could contain unintended errors or omissions.

Chapter Nineteen

The Bainbridges, A Tale of Intrigue and Terror

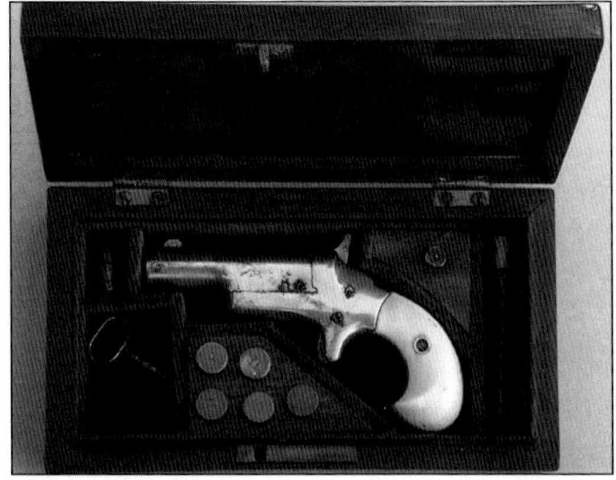

The inscription reads "Wm. E. Bainbridge / Council Bluffs". Bainbridge was a diplomat in China during the Boxer Rebellion.

On January 6, 2015, I purchased an original Colt .41 cal. #3 Thuer Derringer, which had this period inscription presented on the case lid: "William E. Bainbridge, Council Bluffs", Serial #9650. My research on Bainbridge revealed a brilliant, sensitive yet brave, complex persona.

William Elmer Bainbridge was born in the Village of Mifflin, Wisconsin (population 321), New Year's Day, 1861. In 1886, he graduated from the University of Wisconsin and in 1889 from the Law Department of that institution with high honors. He then moved to Council Bluffs, Iowa, in 1891, where he resided until his mysterious and untimely death.

He was married in Council Bluffs to Mary Sims McCarger, a beautiful, socially prominent lady, on January 31, 1894, and

Bainbridge is third from left shown in a diplomatic meeting.

became a partner in the firm of Sims and Bainbridge, where he practiced law. In 1898, he was honored by President William McKinley, who appointed him as Second Secretary of the United States Diplomatic Corps stationed at Peking, China. He and his wife were in Peking before, and during, the infamous horror-filled Boxer Rebellion.

The Diplomatic party arrived at Peking July 7, 1898, and on July 10th, Mr. Bainbridge, as a member of the legation staff, attended an audience given by the Imperial all-powerful Emperor of China. During that time, Mr. Bainbridge was often invited, surprisingly, to audiences granted by the very inaccessible Emperor and Empress Dowager in their private quarters.

Bainbridge and his wife, Mary, on a leisurely outing on horseback in Peking, 1899.

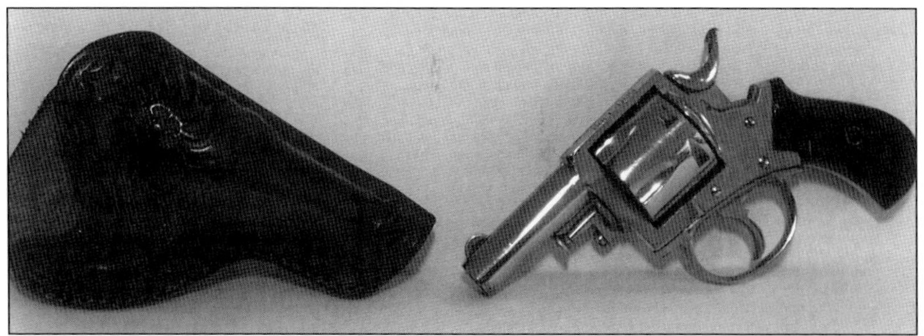

British Bulldog .38 cal. Revolver and Holster

At the time of Bainbridge's arrival in China, the modern thinking emperor was deeply interested in his efforts toward the reformation of his government by the introduction of foreign methods and ideas along many diverse lines. The following September, these efforts ended disastrously in the now famous coup d'état and the resumption of power by the Empress Dowager.

As Secretary, Mr. Bainbridge was in close contact with the development of anti-foreign sentiment in high Chinese official circles, and watched with deep interest and concern the growth of the secret Chinese sect called, "Boxers" as they spread propaganda.

Many Chinese nationals hated foreigners for introducing opium, missionaries, guns, and numerous Western evils into their culture, which is understandable.

At this time, Bainbridge, sensing danger, began carrying his Colt Derringer in his vest pocket and a British Bulldog .38 Caliber Revolver concealed in his holster.

In the spring of 1900, the Boxer Movement became extremely threatening, and events moved from bad to worse until they culminated in the attack on the City of Tientsin and the violent siege of the legations at Peking. Legations in China were the political and business centers of foreign countries engaged in lucrative trade with China. A legation is one step below an embassy.

Great Britain, America, Russia, Germany, France, Japan, Spain, Austria, The Netherlands, Belgium, and Italy had their residential and business offices in a compound surrounded by high walls and a moat for security.

The Boxers destroyed the railway between Peking and Tientsin on June 5, 1900, and after that, the foreign community at Peking was completely cut off from the outside world, causing terror to sweep through the posh, upper class residents. On June 19th, the Chinese Foreign Office notified the Diplomatic Corps and legation representatives they must leave Peking within twenty-four hours. The diplomats and those whom they were trying to protect were already hemmed in and knew any attempt to leave would result in certain massacre the moment they were outside the walls of the city.

On July 20th, German Minister Baron Von Kettler was assassinated by Chinese soldiers on a principal street near the legation headquarters. The terrified foreign community, including their many children, immediately gathered into the British Legation, since it was the most defensible position. Although most of the other legations were protected, the Austrian, Italian, The Netherlands, and Belgian legations were abandoned and immediately burned by the Chinese. Imagine the fear this created. The defensive force consisted of only about 400 marines, who arrived in the nick of time on May 31, 1900.

Mary Bainbridge was known to call the marines "Our Boys" with pride and to treat each of them with love and care.

The Chinese opened fire at 4:00 P.M. on June 20th. The entire section of the city surrounding the legations was burned in an effort to open a path by which they might enter the Legation District and annihilate everyone inside. Rifle and artillery fire was almost constant from June 20th to July 7th, seventeen terrifying days accompanied by cries of death on both sides. By this time, the Bainbridge's and all those under siege were reduced to eating horse meat and enduring multiple horrors, sickness, and death.

Then, having learned of the capture of Tientsin by the allied troops, the Chinese ceased their regular attacks but kept a close watch on the foreign community, ever alert to shoot anyone who exposed himself within rifle range. Many foreign residents in the legations were killed or wounded. As soon as the allies began their march from Tientsin, the attacks increased in violence until the night of August 13th, when the allies were within five miles of Peking. Then the most savage and determined effort was made to overwhelm the legation. This was recognized as their last opportunity to destroy the foreign community. After midnight, a detachment of brave Russian, German, and American troops moved up and attacked one of the city gates. This action drew away many of the Chinese to defend the walls and probably saved those who were still alive.

The allies attacked in force the morning of the 14th and entered the legation, relieving those who had been in deadly peril for over two grueling months. During the siege from May 6 to August 18, 1900, seventy-two legation members were killed and 150 wounded.

This event will go down in history as one of the most remarkable defenses in the annals of warfare. As a member of the Diplomatic Corps, Mr. Bainbridge was responsible for the safety of all. His beloved wife, Mary, was in grave peril as well. The long strain and horrors of that scene for two months very seriously impaired his physical and mental health. After saving about 420 foreign representatives along with the other diplomats and marines, President Theodore Roosevelt appointed Bainbridge Commissioner of the Chinese Claims Proceeding. He was instrumental in adjusting claims on behalf of American citizens for damages caused by the Boxer Rebellion. These damages amounted to over $4 million in gold in 1900. You can imagine the damages in today's dollars.

At this time, Mr. and Mrs. Bainbridge returned to Council Bluffs for a period of rest, recovery and relaxation. His service in Peking was so well respected by all sides that President Roosevelt appointed him American Commissioner to Venezuela. He served brilliantly on the Arbitration Board of Claims against that country following the Venezuelan Civil War and Crisis of 1902-03. His legal opinions were marked by keen, incisive logic and displayed a wide knowledge of international law as well as exceptional people skills.

In 2002, unbelievable information surfaced from classified top secret German archives related to the Venezuelan Crisis of 1902-03. The disputes nearly brought about the German invasion of Long Island, New York, and Washington, D.C., along with a full naval confrontation of the dreadnaught fleets of both the U.S. and Germany. The deceptive, horrific German plans included an attack on Puerto Rico with the purpose of annexing that country into Kaiser Wilhelm II's German Empire. Bainbridge was

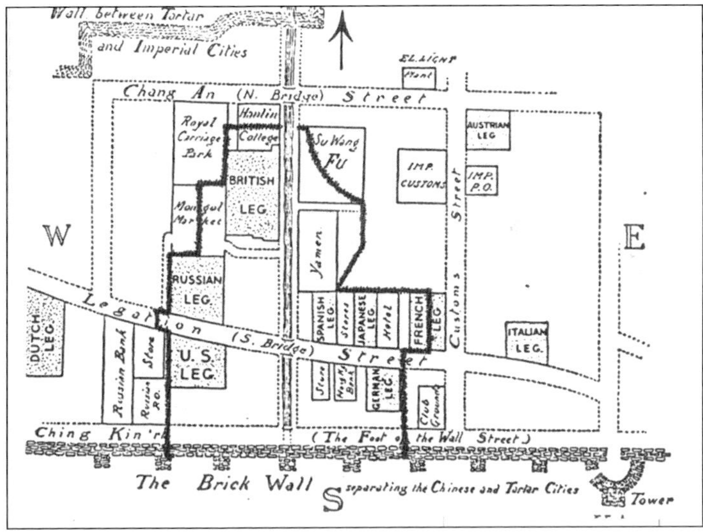

Diagram of the legations during the Boxer Rebellion.

aware and had to deal with this intense hostility from Germany on a daily basis but, of course, did not suspect invasion of the U.S.

In 1906, Bainbridge was appointed Special Agent of the U.S. Treasury Department, overseeing all U.S. Customs Agents in France. This was an enormous and stressful assignment, but the ambitious Bainbridge could not turn this opportunity down. Again, his popular and helpful wife, Mary, accompanied him to Paris, where they leased a beautiful spacious apartment.

Shockingly, Bainbridge took his life with a gunshot by his own hand on April 15, 1909. Was it the horrors of the Boxer Rebellion or the job pressures in Venezuela and Paris, or simply clinical depression? Only God knows!

Six years later, observers believe Mary died of a broken heart. They were buried next to each other in Walnut Hills Cemetery.

Whenever I hold this exquisitely cased Colt Derringer, I know I have history in my hand.

References

The Personal Diary of Mary A. Bainbridge from the State Historical Society of Iowa.

Post Mortem Address About William E. Bainbridge published by the Iowa State Bar Association.

Through the Boxer Siege in Peking published by the Wisconsin University *Alumni Magazine.*

"Executive Journal" of April 21, 1898.

The Venezuelan Arbitration Before the Hague Tribunal, 1903, published in Washington, D. C. by the Government Printing Office.

"Venezuelan Crisis of 1902-1903" on Wikipedia.

"Uncle Sam's War of 1898 and the Origins of Globalization" (discusses Germany's plans to attack the U. S. in 1902-1903).

Chapter Twenty

Rare Cased American Dueling Pistols – A Tantalizing Relationship Between Maker and Owner

Mr. Prosper D. Vallee, at forty-six years of age, was at the peak of his career as gunmaker and hardware/outdoor equipment merchant in 1835. He owned his shop, with residence above, at 101 South Second at Walnut Street, in the heart of downtown Philadelphia. A successful property owner, Vallee, of French descent, was a prominent member of the middle or upper middle class. At the time, only ten percent of Philadelphians owned any real estate, and wealth was extremely skewed. Philadelphia was the second largest, and richest, city in America

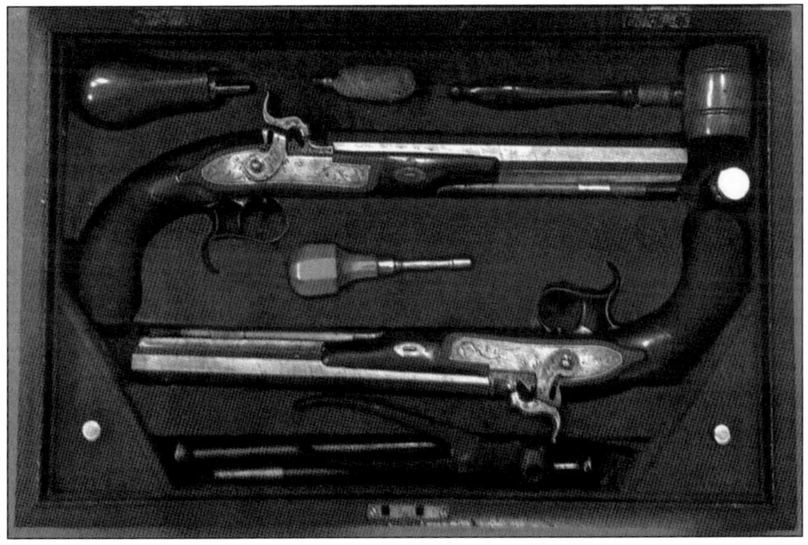

Cased set of pristine, untouched American dueling pistols with all original accessories by Prosper Vallee of Philadelphia.

Engraved lock showing the unusual duelist aiming pistol and P. Vallee's signature.

with about 88,000 residents. The top one percent—or only 880 people—owned fifty percent of the wealth, the top ten percent owned ninety percent of the wealth, and the remaining ninety percent, or 79,200 people owned only ten percent of the wealth, so most citizens were struggling to make ends meet. The poverty line was $518 per year for a typical family. America's wealth in 2020 is much more dispersed, plus we have many entitlements and safety nets.

Prosper was celebrating the 26th birthday of his second wife, Amelie, with whom he had been married for three years. She had recently delivered Jocelyn, their second daughter, in addition to their first daughter, Elise. Prosper's first wife, Ruth, had given him three sons—Elijah, Caleb, and Isaac—and three daughters—Isabella, Sarah, and Katherine. Ruth and baby, Isaac, died in 1832 during the Cholera epidemic, leaving Prosper devastated. Within a year, the lonely Prosper found love again.

By 1836 Amelie and Prosper were happy, busy, and devoted parents. Their home above his shop was large with a beautiful, well-appointed interior consisting of five bedrooms, a parlor, dining room, kitchen, and four fireplaces. They all got a warm weekly bath in the same tub with Amelie and Prosper first, followed by the children in order by age, so one can imagine what the water was like for the youngest child, but this was typical in

1836. The anecdote, "Don't throw the baby out with the bath water" came from this era.

Prosper and his family had been prominent members of the prestigious First Presbyterian Church of Philadelphia since 1830, when Albert Barnes, a handsome charismatic minister from New Jersey, ascended to the office of First Minister and leader of the First Presbyterian Church. The affluent and well-born Reverend Barnes graduated from Princeton Seminary in 1824, and in 1825 was called to the pulpit of First Church in Morristown, New Jersey. There Barnes, a spellbinding evangelist, preached temperance, sexual continence, celebration of industrialism, and acquisitive man.

Prosper Vallee's shop at 101 South Second Street, later John Krider's shop.

In Philadelphia, Reverend Barnes set off a firestorm of controversy in the Quaker Capital of Orthodoxy. The "Old Guard" twice charged him with heresy. Yet, through all the turmoil of the trials, Barnes commanded the unyielding loyalty of his wealthy parishioners; they saw him not as a heretic, but as a prophet of the "New Order." Prior to revivalism of the 1820-1850 period, the "Old Order" colonial American men loved drinking, dancing, brawling, fornicating, gambling, relaxed work schedules, and few rules. Many women joined them in this life, which they saw as good and natural. They saw evangelism as a threat that would take all the fun out of their short and difficult lives. Of course, many women supported evangelism, believing it would lead to a better family life.

The years 1820 to 1850 were tumultuous times; labor unions were fighting employers with strikes of over 20,000 employees;

the ninety percent American-born were fighting the ten percent foreign-born; whites were fighting Asians, blacks, and immigrants; religions were fighting each other; danger was everywhere, and security was nowhere, except for a few fortunate members of the top ten percent.

Painting of Amelie Vallee on her 30th birthday.

As a result, Reverend Barnes and his wife received threats to their lives, family, and property, and felt he may even be challenged to a duel. He let his congregation know of his potential peril from the pulpit. After understanding these threats, Prosper offered the Reverend his finest custom-made pair of dueling pistols worth $50 to $60 for the bargain price of $30, and a well-made rifle for $10. Barnes readily accepted, and in return, invited Prosper to be an elder of the most prestigious church in Philadelphia, a position most members would covet.

Prosper, always a perfectionist, so true to his nature, had the initials of Albert Barnes inscribed on both pistols: AB No. 1 and AB No. 2. Perhaps his perfectionism accounts for his low production of guns, since only a few signed guns are known, according to Stuart Mowbray, editor of *Man At Arms* magazine. In addition, Prosper may have spent more time and effort as a merchant. He is known through newspaper ads to have imported many English and French fowlers for sale, but these he did not likely sign according to Mr. Ronald

AB are the initials of the owner, Albert Barnes

Pennsylvania Antique Arms Collectors Association display of Philadelphia-made guns for the NRA Convention, 1991.

Gabel, an expert on Philadelphia gunmakers. Mr. Gabel agrees with Stuart Mowbray that signed Prosper Vallee guns are scarce, valuable, and interesting. He provided me a photograph of the Pennsylvania Antique Arms Collectors Association display for the NRA Convention in 1991, and said the display featured over eighty Philadelphia gunmakers, with only one pistol signed by Prosper Vallee.

Prosper became a close friend and confidant of Reverend Barnes, and eventually discovered an additional reason why the Reverend may well need protective weapons. Barnes confided that like Saint Paul, he had an affliction he could not rid himself of, a weakness for women, especially members of the congregation with whom he had a counseling relationship. Prosper understood this difficult conflict, and comforted his friend with empathy. There is an unsubstantiated rumor claiming that Barnes did eventually fight a duel with an irate husband of one of the ladies in his flock, but no proof could be found.

Where did Prosper Vallee come from and what contributed to his talent and success? Prosper's parents, Marie and Francois Vallee, immigrated to America in 1779 from St. Henri, Quebec, which was well known for its artisans. His father was a clockmaker and his mother a portrait painter. At some point she painted Prosper as a younger man in a primitive style. They settled in Lancaster, Pennsyl-

First Presbyterian Church of Philadelphia.

vania, where Prosper and his siblings grew up observing their parent's skills and talents, while absorbing a lot of the Quaker culture of hard work, honesty, and wholesome living so dominant in the area. This probably accounts for the numerous Quaker names in the Vallee family.

Prosper found himself fascinated with guns in his early teens and was apprenticed to John Demuth, gun maker in Lancaster, in 1804. At this time, Prosper discovered he had a two-year-old half-brother named Albert G. Bird, whose mother was Rebecca Bird, a young widow in Lancaster. Rebecca was an attractive and talented milliner who made hats for the most fashionable of ladies. Apparently, Francois found Rebecca appealing, perhaps irresist-

Handsome and charismatic Reverend Albert Barnes, ca. 1833.

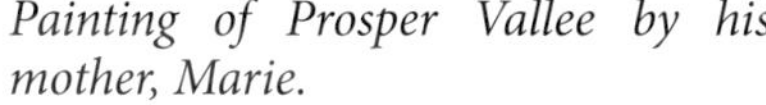

Painting of Prosper Vallee by his mother, Marie.

A portrait of John Krider, probably taken in the late 1870s.

ible, so she became his mistress and in 1802, a mother. This discovery shocked neither Prosper nor Marie; they were aware of the tendency of successful Frenchmen of the day to take a mistress. Prosper looked forward to getting to know little Albert.

By 1819, Prosper had been moving from apprentice, to journeyman, to master of the art of gunmaking, so having saved his money, in 1820 he moved to Philadelphia and opened his own gun shop at 101 S. Second Street at Walnut. He soon achieved enough success that in 1826 he apprenticed a very talented man, John Krider. The business expanded rapidly in scope and size, taking on additional outdoor equipment for miners, fishermen, explorers, farmers, archers, and others. About 1831, having completed his gunmaking and merchandising apprenticeship, John Krider left Vallee to set up his own shop. Vallee took additional apprentices and continued to grow his business. His half-brother had opened his own engraving company in Philadelphia at 100 Union Street, and had subcontracted numerous engraving jobs for Prosper Vallee.

All was going well until 1837, when a deep financial disaster struck America and would last six long years. The panic idled thousands of employees and destroyed or debilitated unions and cooperatives; wage earners had to take whatever they could find. Few emerged unscathed. The years 1837 to 1843 saw many bank failures, and suspension of specie (gold and silver) payouts to their depositors. By 1842, streets were deserted, large homes closed up for rent or sale; no business, no money, no hope, and no confidence existed. Auctions sold people's belongings for twenty-five cents on the dollar. The rich saw their investments depreciate. There was deflation of wages and prices. Martin Van Buren was the U.S. president and took much of the blame, but the underlying causes were a real estate bubble, an erratic American banking policy, reckless acquisitions and general overspending, and finally, capital was dear and scarce.

Prosper did not escape this meltdown, because he was overextended and was forced to downsize his operations and move to 1 Locust Street, a less expensive location. Ironically, John Krider, his former apprentice, bought Prosper's building and business cheap and went on to become a legendary success in the gun and sportsman's goods business until his death in 1886.

By 1845, Prosper wanted a better location, so he moved to 173 S. Second Street, only a block away from his original business. He was discouraged and worried about the wellbeing of his wife and two teenage daughters. In 1847, Prosper moved his family to Brooklyn, New York, in hopes of a better future. He eventually found a middle managers job as an employee of the Schuyler, Hartley, and Graham Company, the largest firearms and sporting goods wholesaler in America at the time. He was comfortable and able to provide a good life for his family.

Prosper died at 71 years of age and was buried in Greenwood Cemetery in Brooklyn on May 16, 1861. Amelie was buried next to him November 21, 1889. Fortunately, you can see for yourself the expertise and artistic ability Prosper Vallee was able to incorporate into his handmade firearms at the end of the pre-industrial period!

References

Mr. Stuart Mowbray, Editor, *Man at Arms* magazine and noted collector and expert on dueling pistols.

The book, *Working People of Philadelphia 1800 to 1850*, by Bruce Laurie.

The Historical Society of Pennsylvania—Mr. David Haugaard, Director, performed ten hours of research on my behalf.

Mr. Ronald G. Gabel, expert on Philadelphia guns and gunmakers from Slatington, Pennsylvania.

Mr. Charles Layson, owner of Antique and Modern Firearms, Lexington, Kentucky.

U.S. Census Records and U.S. Inflation Records.

Prosper Vallee's Ledger of Customers and Day Book from the Historical Society of Pennsylvania.

John Krider's internet website.

Afterword

For me, it has been a gratifying experience to bring you these true stories of fascinating people and unique firearms. Each life depicted in these tales would have remained buried in the dust of history if it had not been for these identified handguns that brought their stories to life. Mr. John G. Hamilton carried out an in-depth study of identified pistols and revolvers in 1963 with some very interesting findings, as follows:

1. Most presentation, inscribed, or identified arms occurred just before, during, and after the Civil War, after which the custom steadily declined.
2. Over ninety percent of all inscriptions are found on Colt pistols. The most inscribed non-Colt revolver is the Smith and Wesson, #2 Army model.
3. The Colt revolver most often used for presentation and inscriptions is the l851 Navy. Second is the 1849 Colt pocket model, and the third is the Model 1862 Colt police.
4. Most inscriptions are found on the back straps of pistols, but some are found on the butt strap, trigger guard, barrel, grips, or an inscription may be carved into the grips.

It has taken over 1,000 hours to research and write these twenty true accounts. However, the personal satisfaction is hard for me to quantify. The history of our dynamic America, my love for special antique pistols, and my enjoyment of sharing these stories is a far greater reward than I ever imagined. Hunting and sport shooting is another aspect of guns that I hold dear. We citizens of this great land must protect our second amendment rights, support the

NRA, and do all we can to safeguard our God-given right to keep and bear arms, for enjoyment and protection of families.

Since the majority of my stories involve the Civil War, I think you would find these statistics very interesting (numbers are best approximations):

Union Army enlistments —	2,750,000
Confederate Army enlistments —	800,000
Free Black soldiers —	178,000
American Indian soldiers —	3,500
Union Army average age —	26 years

127 Yankee soldiers were only thirteen years old, and 2,300 were over 50.

Average weight was about 139 pounds and height was 5'8".

The Union soldiers were organized into 3,550 different units:

Infantry Regiments —	2,139
Cavalry Regiments —	270
Heavy Artillery —	61
Engineers —	13
Light Infantry Battalions —	9
Artillery Batteries —	430

Confederate soldiers were organized into about 1,526 units:

Infantry Regiments —	640
Cavalry Regiments —	139
Artillery Regiments —	16
Artillery Batteries —	228

The Northern Army had deaths of 360,250. Confederates lost 259,000 lives, for a total of about 620,000 dead Americans.

About 195,000 Yankee soldiers were put in southern prisons, of which 30,200 died in captivity.

Approximately 215,000 Confederates were captured and imprisoned, of which 26,000 died in captivity.

The Union had 583 Generals, while the Confederates had 425.

Where the fighting took place in terms of numbers of events from battles to skirmishes to sieges etc:

Virginia	2,150
Tennessee	1,468
Missouri	1,160
Mississippi	778
Arkansas	772
West Virginia	634
Louisiana	568
Georgia	549
Kentucky	454
Alabama	338
North Carolina	314
South Carolina	238
Maryland	204
Florida	166
Texas	92
Indian Territory	88
California	87
New Mexico Territory	76

The greatest number of soldiers brought together in the Western hemisphere occurred on December 13, 1862 at Fredericksburg, Virginia. About 200,000 soldiers participated resulting in a bloody defeat for the union.

Gettysburg, Pennsylvania was the scene of the bloodiest battle on July 1-3, 1863, where 5,800 died and over 43,000 were wounded.

Spies flourished, with both the North and the South employing about 1,000 spies each. Of these, about 80 to 100 were executed.

Naval power at the peak for the Union found the North with approximately 3,450 guns mounted on ships, which were able to cripple the South with blockades and other actions.

The Confederate Navy was weaker, never having more than 650 to 700 guns on ship decks. This lack of Naval power cost the South dearly.

As you readers know by now, I am a strong advocate for our second amendment rights as well as a supporter of our entire constitution. A particularly onerous situation exists when our government continually attacks our right to keep and bear arms because the politicians say we will all be safer. The true statistics are as close as your computer, some of which I will enumerate here:

Since Great Britain passed its comprehensive ban on handguns, criminal use of firearms has increase by forty percent to 3,685 incidents from 2,648.

Australia confiscated 640,000 pump action shotguns and rifles in addition to semi-automatics at a cost of half a billion dollars of the taxpayer's money.

Australia's results to date:

Armed robbery - up sixty-nine percent

Assaults - up twenty-eight percent

Gun murders - up nineteen percent

Home invasions - up twenty-one percent

Washington D.C. has 1,508 violent crimes per 100,000 population, making it the most dangerous place in the USA, yet, until recently, it had the strictest gun laws in America.

The second most dangerous place in the U.S. is South Carolina, with 766 violent crimes per 100,000 people, almost half the crime of Washington, D.C.

Since the outset of the Chicago handgun ban, the percentage of Chicago murders committed with handguns has averaged forty percent higher than it was before the law took effect.

Since the outset of the Florida Right-to-Carry law, the Florida murder rate has averaged thirty-six percent lower than it was before this law took effect, while the U.S. murder rate has averaged fifteen percent lower.

In Texas, since the outset of their Right-to-Carry law, the Texas murder rate has average thirty percent lower than it was before the law took effect.

About 10,800 murders are committed in the U.S. using firearms each year. And yet, based on a study by the *Journal of Quantitative Criminology*, U.S. civilians use guns to defend themselves and others from crime about 1,000,000 times per year. What would have happened in the above million events if the protector had no firearms? The truth is we would all be safer if every household had a gun and someone trained to use it like Switzerland, which has one of the lowest crime rates in the world.

In summary, it is so easy to get the true facts and statistics on line; one has to wonder what motivates these anti-gun politicians?

Credits

The short stories in this book were previously published in various magazines that have given me permission to use them as follows.

Man at Arms Magazine

1. The Lefaucheus revolver of John Wysong – Vol. 28, No. 5, October 2006.
2. A presentation Colt M 1851 NAVY Vol. 20, No. 3, June 1998.
3. Duels, Honor, and The Confederacy – Major General John C. Breckinridge's Horse Pistols.

The Rampant Colt Magazine

1. An early California Sheriff and his inscribed Colt M 1849 Pocket Model – Spring 2005
2. From affluence to prison for Sergeant Al D. Fobes of the famous 11th Indiana volunteers – spring 2009.
3. The military life of a brawling Indian fighter and Civil War hero – spring 2011.

Remington Society Journal

Private App's N. M. Remington Revolver – 3rd qtr 2001.

Civil War Historian

1. John Hancock, Civil War veteran, March-April 2007.
2. Major Robert McClure, his Colt, and Sherman's March to the sea – Vol. 1, issue 6, November 2008.

The Gun Report

1. The guns of Caldwell, Kansas Vol. 49, Number 1, June 2003.
2. Engraved and inscribed Smith & Wesson revolver leads to discovery of a sinister plot – Vol. 53, No. 7, December 2007.
3. One American family provides four soldiers to fight in the great war of the American Rebellion-Vol. 54, No. 9, February 2009.

Smith & Wesson Collectors Journal

1. John H. Masonheimer – A Civil War hero who got a second chance. Vol. 38, Number 2, summer 2004.
2. My gosh – what a coincidence – Vol. 35, No. 1, Spring 2001.
3. The Fighting Preacher of the Confederacy – Vol. 46, No. 1, spring 2012

Blue & Gray Magazine

Four very rare inscribed or identified Civil War revolvers – Vol. XXV, #4.

North South Traders Civil War

Confederate Colt M1860 Army shine's light on heroic Virginia family – fall 2013.

About the Author

Jerry W. Pitstick has been a lifelong gun collector interested in antique pistols, sport shooting, and hunting, as well as history in general. He joined the Ohio Gun Collectors Association in 1959 and is a member of the Colt Collectors Association, The Remington Society of America, and the Smith & Wesson Collectors Association. Over the years, Jerry has enjoyed exhibiting his gun collection at display shows and has won several important awards. Researching and writing the history of identified Civil War and Win-the-West pistols and revolvers has been Jerry's main interest since 1998.

"Guns That Talk — History in Your Hand" was the title of Jerry's display at the Ohio Gun Collectors Association in May 1999, which won first place and a $3,000 award.

In October 1999, he competed at Louisville, Kentucky where Ron Dickson's *National Gun Day – The Great Eastern Display Expo* was held. Again, he was honored to win second place against tough competitors. Over the last decade he has participated in over ten competitive display shows, with considerable success.

In 2001, Charles Worman, consultant to the Hubbard Museum, asked to borrow eight of Jerry's guns for a six month exhibit at the Hubbard Museum of the American West in New Mexico. This renowned museum put on an extravaganza of the westward migration that included over 500 important guns organized by time periods from 1803 to 1900. Tens of thousands turned out to see this important exposition from May 24, 2001 to November 24, 2001. After this, he was invited to put on two more museum exhibits—one for Smith & Wesson in Springfield, Massachusetts, and one at Camp Dennison in Cincinnati, Ohio.

Jerry enjoys writing articles for gun and Civil War magazines, including *Man at Arms Magazine, Blue and Gray Magazine, The Gun Report, The Smith & Wesson Collector's Journal, Civil War Historian, North South Traders Civil War, The Remington Society of America Magazine* and *The Rampant Colt.*

Index